Synthesis Lectures on Sustainable Development

Series Editor

Thomas Siller, Civil and Environmental Engineering, Colorado State University, Fort Collins, USA

This series publishes short books related to sustainable development practices relevant to engineers, technologists, managers, educators, and policy makers. The books are organized around the United Nations Sustainable Development Goals for 2015–2030. Design for sustainability along with the economics of sustainable development will be the common themes for each book. Topics to be covered will span all the major engineering disciplines along with the natural and environmental sciences that contribute to an understanding of sustainable development. The goal of this series is to make theory and research accessible to practitioners working on sustainable development efforts.

Shaheemath Suhara K. K. · A. K. Haghi

GIS in Environmental Engineering

Core Concepts for Sustainable Development

Shaheemath Suhara K. K.
Kelappaji College of Agricultural Engineering
and Food Technology
Kerala Agricultural University
Tavanur, Malappuram, Kerala, India

A. K. Haghi
Institute of Molecular Chemistry
Coimbra University
Coimbra, Portugal

ISSN 2637-7675   ISSN 2637-7691 (electronic)
Synthesis Lectures on Sustainable Development
ISBN 978-3-031-79125-3   ISBN 978-3-031-79126-0 (eBook)
https://doi.org/10.1007/978-3-031-79126-0

© The Editor(s) (if applicable) and The Author(s), under exclusive license to Springer Nature Switzerland AG 2025

This work is subject to copyright. All rights are solely and exclusively licensed by the Publisher, whether the whole or part of the material is concerned, specifically the rights of translation, reprinting, reuse of illustrations, recitation, broadcasting, reproduction on microfilms or in any other physical way, and transmission or information storage and retrieval, electronic adaptation, computer software, or by similar or dissimilar methodology now known or hereafter developed.
The use of general descriptive names, registered names, trademarks, service marks, etc. in this publication does not imply, even in the absence of a specific statement, that such names are exempt from the relevant protective laws and regulations and therefore free for general use.
The publisher, the authors and the editors are safe to assume that the advice and information in this book are believed to be true and accurate at the date of publication. Neither the publisher nor the authors or the editors give a warranty, expressed or implied, with respect to the material contained herein or for any errors or omissions that may have been made. The publisher remains neutral with regard to jurisdictional claims in published maps and institutional affiliations.

This Springer imprint is published by the registered company Springer Nature Switzerland AG
The registered company address is: Gewerbestrasse 11, 6330 Cham, Switzerland

If disposing of this product, please recycle the paper.

Preface

This book is aimed to be a reference of theory and practice for the professionals in topographic mapping as it has been evolved into a modern field called geospatial information science and technology to provide practical applications of topographic mapping (environment, geology, geography, cartography, engineering, geotechnical, agriculture, forestry, etc.).

It covers the most recent case studies, programs, maps, graphics, and models of how geospatial information science and technology can be applied to automate mapping processes, collect, process, edit, store, manage, and share datasets, statistically analyze data, model, and visualize large datasets to understand patterns, trends, and relationships to make educated decisions.

This book

- Identifies effective GIS techniques for the management of complex infrastructure.
- Discusses the capabilities, requirements, and limitations of available GIS applications.
- Provides advanced GIS database techniques.
- Plays a remarkable role in practical hydrological, topological, and cartographic modeling.
- The topics included in this book on GIS are of utmost significance and bound to provide incredible insights to postgraduate students.

The book focuses on using GIS and innovative models to solve real environmental issues with analytical outputs and covers advanced modeling along with advanced data collection or visualization approaches.

The readers also learn the real-world applications of remote sensing and GIS in disaster management.

Kerala, India Shaheemath Suhara K. K.
Coimbra, Portugal A. K. Haghi

Contents

Applications of GIS and Remote Sensing in Hydrology, Water Resource Management and Sustainable Development 1
1 Overview 1
2 Remote Sensing 2
  2.1 Science Behind Remote Sensing 2
  2.2 Active and Passive Remote Sensing 4
3 Different Remote Sensing Data 6
4 Characteristics of Remote Sensing Data 6
  4.1 False and Natural Colour Composites 6
  4.2 Image Resolution 12
  4.3 Spatial Resolution 12
  4.4 Temporal Resolution 13
  4.5 Spectral Resolution 13
  4.6 Vector and Raster Types 14
5 Various Kinds of Remote Sensing Data 16
  5.1 Digital Elevation Models (DEMs) 16
  5.2 Digital Terrain Models (DTMs) 18
  5.3 Multispectral Aerial Imageries 18
  5.4 Administrative Boundary Maps 20
  5.5 Meteorological Data 21
6 Geographic Information Systems (GIS) 21
7 Integration of GIS and Remote Sensing 23
8 Conclusion 24
References 24

Land Use Land Cover Change Detection Using Google Earth Engine 27
1 Introduction 27
2 History of LULC Change Detection with RS, GIS, and ML 28

3	Machine Learning in LULC Classification	29
	3.1 Supervised ML Algorithms	29
	3.2 Unsupervised ML Algorithms for LULC Classification	36
4	Processes Involved in the LULC Classification Using Google Earth Engine	37
5	Case Study	37
	5.1 Description of the Study Area—Upper Bhavani Basin	37
	5.2 Methodology Adopted	39
6	Results and Discussion	41
	6.1 Trends and Observations	42
	6.2 Accuracy Assessment	44
7	Conclusion	46
References		47

Watershed Prioritization for Rainwater Harvesting Using Multi-criteria Analysis, GIS, and Remote Sensing ... 49

1	Introduction	49
2	Materials and Methods	51
	2.1 Study Area—Noyyal River Basin	51
	2.2 Data Collection	52
	2.3 Data Integration and Analysis in GIS Environment	57
3	Results and Discussion	57
	3.1 Slope Map	58
	3.2 Drainage Density Map	59
	3.3 Soil Map	60
	3.4 LULC Map	60
	3.5 Rainfall	61
	3.6 Geology Map	62
	3.7 Geomorphology Map	63
	3.8 Lineament Density	64
	3.9 Groundwater Potential Zone Map	65
4	Conclusion	67
References		67

Study on Rainfall Variability and Distribution in Lower Bhavani River Basin, Tamil Nadu 69
1 Introduction 69
2 Study Area—Lower Bhavani 70
3 Rain Gauge Stations 71
4 Rainfall Distribution in the Lower Bhavani Basin 73
5 Methodology for Estimating Average Annual, Seasonal, and Decadal Rainfall Using Daily Rainfall Data 73
6 Results and Discussion 74
   6.1 Temporal Variations in Annual Rainfall Distribution in Lower Bhavani Basin 75
   6.2 Seasonal Distribution of Rainfall 77
   6.3 Spatial Analysis of Seasonal Rainfall Distribution 81
   6.4 Decade Wise Rainfall Distribution 82
7 Conclusion 88
References 89

Geomorphometric Analysis of Nileswar Sub-Watershed, Kerala Using GIS and Remote Sensing 91
1 Introduction 92
2 Materials and Methods 93
   2.1 Description of the Study Area 93
   2.2 Methodology 93
3 Results and Discussion 100
   3.1 Linear Aspects of Drainage Basin 101
   3.2 Areal Properties of Drainage Basin 107
   3.3 Relief Aspects of Channel Network 109
   3.4 Watershed Prioritization 110
4 Conclusion 112
References 112

About the Authors

Shaheemath Suhara K. K. is an Assistant Professor in the Department of Soil and Water Conservation Engineering at Kelappaji College of Agricultural Engineering and Food Technology, Tavanur, under Kerala Agricultural University. She completed her Ph.D. at Tamil Nadu Agricultural University, Coimbatore. Her research interests focus on conserving natural resources to enhance food and water security. She has research experience in hydrology, geoinformatics (GIS), remote sensing, agricultural engineering, precision farming, irrigation, drainage engineering, and spatial analysis of natural resources. Additionally, she is the developer of the drip design software, Drip Design Check.

A. K. Haghi is a retired professor and has written, co-written, edited, or co-edited more than 1000 publications, including books, book chapters, and papers in refereed journals with over 4200 citations and h-index of 34, according to the Google Scholar database. Professor Haghi holds a B.Sc. in urban and environmental engineering from the University of North Carolina (USA) and holds two M.Sc. degrees, one in mechanical engineering from North Carolina State University (USA) and another one in applied mechanics, acoustics, and materials from the Université de Technologie de Compiègne (France). He was awarded a Ph.D. in engineering sciences at Université de Franche-Comté (France). He is a regular reviewer of leading international journals.

Applications of GIS and Remote Sensing in Hydrology, Water Resource Management and Sustainable Development

Abstract

This chapter introduces the fundamental concepts of GIS and Remote Sensing, their integration into hydrology, and their role in addressing modern-day water resource challenges and sustainability goals. This chapter explores the applications of GIS and Remote Sensing in hydrology, including monitoring surface water bodies, soil moisture, and vegetation health. The chapter also discusses the role of these technologies in sustainability, including the identification of sustainable water sources, assessment of environmental impact, and development of conservation strategies.

Keywords

GIS • Remote sensing • Raster data • Landsat

1 Overview

In the rapidly evolving field of environmental science, Geographic Information Systems (GIS) and Remote Sensing (RS) have emerged as vital tools, particularly in hydrology, water resource management, and sustainability. The application of GIS and RS in hydrology, water resource management, and sustainability represents a significant advancement in environmental science and technology [1]. As the global community continues to face water-related challenges, the integration of GIS and RS will be crucial in developing sustainable solutions that ensure the availability and quality of water resources for future generations. These technologies enable the collection, analysis, and interpretation of spatial and temporal data, providing detailed analysis on water-related phenomena. Their application spans from understanding hydrological processes to managing water resources efficiently and promoting sustainable practices. This chapter introduces the fundamental concepts of GIS and remote sensing, their integration into hydrology, and their role in addressing modern-day water resource challenges and sustainability goals.

Effective water resource management aims to balance the competing demands of water use, ensuring availability for drinking, agriculture, industry, and ecosystem health. GIS and RS are vital in this regard, providing the data needed to model water systems, monitor usage patterns, and detect anomalies such as droughts or contamination with ease [2].

GIS and RS technologies have transformed the way we study and manage water resources. They offer incomparable proficiencies in monitoring, analysing, and visualizing natural resources data across various scales, from local watersheds to global water cycles. The integration of these technologies allows for a more detailed understanding of the dynamic and complex nature of hydrological systems. This, in turn, supports the development of effective water management strategies that are essential in a world, facing increasing water scarcity, climate variability, and environmental degradation [2, 3].

Remote Sensing is particularly valuable in hydrology for monitoring surface water bodies, soil moisture, topography and vegetation health. Sustainability in water resource management entails meeting present water needs without compromising the ability of future generations to meet their own. This requires an integrated approach that considers environmental, economic, and social factors. GIS and RS contribute to sustainability by offering tools for detailed planning and monitoring [4].

These technologies enable the identification of sustainable water sources, assessment of the environmental impact of water use, and development of conservation strategies. For example, GIS can help design efficient irrigation systems, while RS can monitor the health of watersheds and the effects of climate change on water resources.

2 Remote Sensing

Remote sensing is the acquisition of information about the Earth's surface through the use of sensors that are not in direct physical contact with the object or area being observed. These sensors are mounted on various remote sensing platforms, such as satellites, aircraft or unmanned aerial vehicles (UAVs) like drones.

2.1 Science Behind Remote Sensing

The science behind remote sensing involves the use of various technologies to collect and analyse data about the earth's surface, including its physical properties, composition and changes over time. This technology captures data across various spectral bands, revealing details that are often invisible to the naked eye.

2 Remote Sensing

Remote sensing is an interdisciplinary field that draws on a range of scientific disciplines to collect and analyse data about the earth's surface. Here are the sciences behind remote sensing and their fundamental aspects.

a. *Electromagnetic Theory*

Electromagnetic theory is the foundation of remote sensing. It explains how electromagnetic radiation interacts with the earth's surface. *Electromagnetic radiation* is a form of energy that travels in the form of waves, including visible light, infrared, and microwave radiation. Remote sensing sensors detect and measure the radiation that is reflected, emitted, or transmitted by the earth's surface [5].

b. *Optics*

Optics is the study of the behaviour of light and its interactions with matter. In remote sensing, optics is vital for understanding how light interacts with the Earth's surface. Reflection, refraction, and absorption are key concepts in optics that help explain how sensors detect radiation. For example, reflection occurs when light bounces off a surface, while absorption occurs when light is absorbed by a material [5].

c. *Spectroscopy*

Spectroscopy is the study of the interaction between matter and electromagnetic radiation. In remote sensing, spectroscopy is used to analyse the spectral signatures of different materials. A spectral signature is a unique pattern of radiation that is reflected, emitted, or transmitted by a material [5].

One of the key concepts in RS is the idea of spectral reflectance, which refers to the amount of radiation that is reflected by an object or surface at different wavelengths. Spectral reflectance is important because it provides a way to identify and distinguish between different objects or surfaces based on their unique spectral signatures. By analysing the spectral reflectance of an object or surface, one can infer its composition, texture, and other properties. For example, vegetation has a unique spectral reflectance signature, with high reflectance in the near-infrared (NIR) and low reflectance in the red and blue parts of the visible spectrum. This is because vegetation contains chlorophyll, which absorbs radiation in the red and blue parts of the spectrum and reflects radiation in the NIR. Water has a distinct spectral reflectance signature, with high reflectance in the blue and green parts of the visible spectrum and low reflectance in the NIR. Soil has a variable spectral reflectance signature, depending on its characteristics such as moisture content. For example, dry soil tends to have higher reflectance in the visible spectrum, while wet soil tends to have higher reflectance in the NIR [5].

d. *Radiometry*

Radiometry is the study of the measurement of electromagnetic radiation. Radiance and irradiance are key concepts in radiometry that describe the amount of radiation that is emitted or reflected by a surface. In remote sensing, radiometry is used to quantify the amount of radiation that is detected by sensors [5].

e. *Atmospheric Science*

Atmospheric science is the study of the Earth's atmosphere. In remote sensing, atmospheric science is crucial for understanding how the atmosphere interacts with electromagnetic radiation. The atmosphere can absorb, scatter, or transmit radiation, which affects the data collected by remote sensing sensors.

f. *Geophysics*

Geophysics is the study of the Earth's internal and external physical processes. In remote sensing, geophysics is used to understand the geophysical properties of the Earth's surface, such as temperature, moisture, and topography. Remote sensing sensors can detect these properties by measuring the radiation that is emitted or reflected by the Earth's surface [5].

g. *Computer Science*

Computer science is the study of the theory, design, and implementation of computer systems. In remote sensing, computer science is used to process and analyse the large amounts of data collected by sensors. This includes image processing, data compression, and machine learning algorithms.

h. *Cartography*

Cartography is the study of the creation and use of maps. In remote sensing, cartography is used to analyse and interpret the spatial data collected by sensors. This includes understanding the relationships between different features on the Earth's surface and creating maps that visualize the data [6].

2.2 Active and Passive Remote Sensing

There are two main types of remote sensing: active and passive (Fig. 1). Active remote sensing involves the use of sensors that emit their own energy, such as radar and LiDAR,

2 Remote Sensing

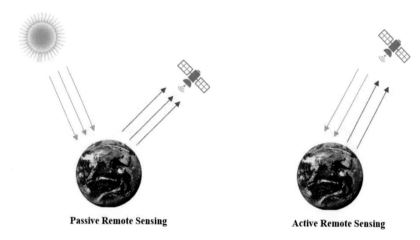

Fig. 1 Schematic of active and passive remote sensing

to illuminate the target area. The sensor then measures the energy that is reflected or scattered back from the target [7]. This type of remote sensing is useful for applications such as:

- Creating high-resolution images of the Earth's surface
- Monitoring land deformation and changes in land use
- Tracking weather patterns and storms

Passive remote sensing, on the other hand, relies on the natural energy emitted or reflected by the Earth's surface, such as sunlight. Passive sensors, like radiometers and spectrometers, measure the energy that is reflected or emitted by the target. This type of remote sensing is useful for applications such as:

- Monitoring crop health and growth
- Tracking changes in ocean currents and temperature
- Studying the Earth's climate and weather patterns

The key differences between active and passive remote sensing are:

- Active remote sensing uses its own energy source, while passive remote sensing relies on natural energy.
- Active remote sensing can be used day or night, regardless of weather conditions, while passive remote sensing requires sunlight and clear skies.

3 Different Remote Sensing Data

There are different types of remote sensing, including optical sensing, multispectral sensing, hyperspectral sensing, thermal sensing, and radar sensing. Each type of sensing has its own strengths and weaknesses, and is used for specific applications. Optical sensing, for example, captures electromagnetic radiation in the visible, near-infrared, and thermal infrared regions of the electromagnetic spectrum. This type of sensing provides information about crop health, vegetation indices, and land cover. Multispectral sensing, on the other hand, captures data in several discrete bands within the electromagnetic spectrum. This type of sensing enables the analysis of specific wavelengths relevant to vegetation health and crop monitoring [8]. The details regarding each type of remote sensing are summarized in Table 1.

Remote sensing data involves a diverse range of formats, each catering to specific applications and requirements. Some of the major forms of remote sensing data include: digital elevation models (DEMs), digital terrain models, multispectral aerial imageries, administrative boundary maps, and more.

4 Characteristics of Remote Sensing Data

These remote sensing data differ in several aspects, such as:

- False and natural colour composites
- Image resolution
- Spatial resolution
- Temporal resolution
- Spectral resolution
- Vector and raster data

4.1 False and Natural Colour Composites

Earth-observing satellites are equipped with multispectral sensors that measure the electromagnetic radiation reflected or emitted from the Earth's surface. These sensors capture data across various regions of the electromagnetic spectrum, including visible light, near-infrared, and shortwave infrared. The range of wavelengths measured by a sensor is referred to as a band, which can be displayed as a grayscale image or combined with other bands to create a colour composite image [12].

There are two primary types of colour composites: natural colour composites and false colour composites. The way the image is represented, with false colour composites using

Table 1 Remote sensing technologies: characteristics, applications and data sources

Type of remote sensing	Description	Strengths	Weaknesses	Applications	Data sources
Optical remote sensing [9]	Uses visible and NIR to detect the reflected radiation from the earth's surface	High spatial resolution, low cost, widely available	Affected by atmospheric interference Limited to day time and clear weather	LULC mapping, urban planning	Landsat 8, Sentinel-2, Planet Labs, DigitalGlobe
Multispectral remote sensing [10]	Measures reflected radiation in multiple spectral bands (typically 3–10 bands)	Improved spectral resolution, can distinguish between different land cover types	Lower spatial resolution than optical sensing, still affected by atmospheric interference	Crop monitoring, land cover classification, environmental monitoring	Landsat 8, Sentinel-2, MODIS, ASTER
Hyperspectral sensing [11]	Measures reflected radiation in hundreds of spectral bands	High spectral resolution, can identify specific materials and chemicals	High cost, complex data analysis, affected by atmospheric interference	Mineral exploration, crop stress detection, environmental monitoring	Hyperion, AVIRIS, HyMap, PRISMA
Thermal sensing	Measures emitted radiation in the thermal infrared spectrum	Can detect temperature differences, day/night and all-weather capability	Lower spatial resolution, affected by atmospheric interference	Land surface temperature (LST) monitoring, crop water stress detection, wildfire detection	Landsat 8, ASTER, MODIS, VIIRS
Radar sensing [11]	Uses microwave radiation to detect reflected radiation from the Earth's surface	Day/night and all-weather capability, can penetrate vegetation and clouds	Lower spatial resolution, affected by topography and surface roughness	Land deformation monitoring, crop monitoring, disaster response	Sentinel-1, TerraSAR-X, RADARSAT-2, ALOS-2

Fig. 2 Natural colour composites

different wavelengths to highlight specific features, and true colour composites displaying the image in its natural colours.

Natural colour composites display visible red, green, and blue bands, resulting in an image that closely resembles what the human eye would perceive (Fig. 2). This type of composite is often preferred because it appears natural to our eyes, but it can be limited in its ability to highlight subtle differences in features.

False colour composites, on the other hand, utilize bands beyond the visible spectrum to create an image. This allows for the visualization of wavelengths that are invisible to the human eye, such as near-infrared. False colour composites can enhance the spectral separation and interpretability of the data, making it easier to distinguish between different features. There are numerous false colour composites, each designed to highlight specific features or characteristics (Figs. 3, 4 and 5).

The band combination of 7, 6, 4 is a common false colour composite used for urban planning and development (Fig. 4). Here is a breakdown of what each band represents and how to distinguish features in this combination:

Band 7 (Shortwave Infrared 2.2 µm): This band is sensitive to moisture and temperature. It highlights features with high thermal activity, such as:

4 Characteristics of Remote Sensing Data

Fig. 3 Image visualization with different colour composites (bands 4-3-2)

- Urban areas: Appear bright white or light gray due to the high thermal activity from buildings, roads, and human activity.
- Vegetation: Appears dark blue or black, as it has low thermal activity and high moisture content.

Band 6 (Shortwave Infrared 1.6 μm): This band is also sensitive to moisture and temperature, but with a slightly different response. It helps to:

- Distinguish between different types of vegetation: Healthy vegetation appears bright green, while stressed or dry vegetation appears more yellow or orange.
- Highlight soil moisture: Wet soils appear brighter than dry soils.

Band 4 (Red 0.65 μm): This band is sensitive to vegetation health and biomass. It helps to:

- Distinguish between vegetation and non-vegetation: Vegetation appears bright green, while non-vegetation features like roads, buildings, and bare soil appear more reddish or brown.

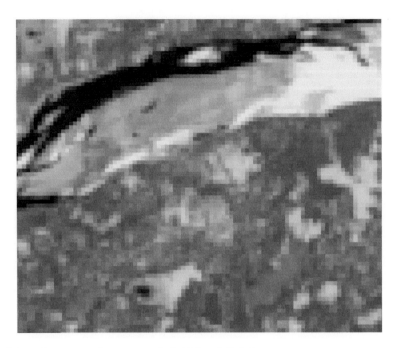

Fig. 4 False colour composites—example (with bands 7-6-4)

In this 7, 6, 4 band combination, features can be distinguished as follows:

1. **Urban areas**: Bright white or light gray (high thermal activity) with a reddish tint (from Band 4).
2. **Vegetation**: Healthy vegetation appears bright green, while stressed or dry vegetation appears more yellow or orange.
3. **Water bodies**: Appear dark blue or black, with low thermal activity and high moisture content.
4. **Soil**: Wet soils appear brighter than dry soils, with a more yellowish or brownish tint.
5. **Roads and infrastructure**: Appear reddish or brownish, with a mix of thermal activity and vegetation signals.

By combining these bands, you can gain insights into urban planning, land use, and environmental monitoring (Fig. 4).

Similarly, the false colour composite (FCC) of Landsat 8 bands 5, 4, and 3 is also often used to enhance the visibility of certain features in the image. In this combination, band 5 is assigned to the red channel, band 4 is assigned to the green channel, and band 3 is assigned to the blue channel. This FCC is useful for identifying vegetation, water bodies, and urban areas (Fig. 5).

4 Characteristics of Remote Sensing Data

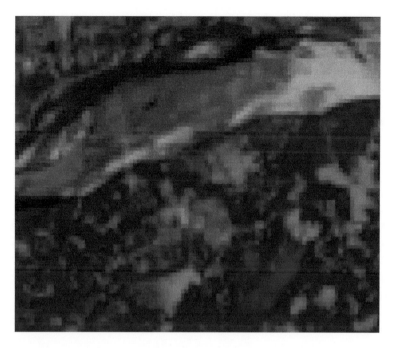

Fig. 5 False colour composites—example (with bands 5-4-3)

Here are some tips for identifying features using the false colour composite of Landsat 8's bands 5, 4, and 3:

1. **Vegetation**: Healthy vegetation typically appears in shades of red or magenta. This is because vegetation reflects near-infrared light strongly, and Band 5 captures this reflection.
2. **Water bodies**: Water bodies appear in shades of blue or black. This is because water absorbs most of the visible light and reflects very little.
3. **Urban areas**: Urban areas appear in shades of gray or cyan. This is because urban areas are often composed of materials that reflect a mix of visible and near-infrared light.
4. **Bare soil**: Bare soil appears in shades of brown or orange. This is because soil reflects a mix of visible and near-infrared light, but not as strongly as vegetation.

Similarly, the band combination of 7, 4, 2 and 2, 4, 7 were used most commonly for lithological identification, especially for sedimentary rock formations [13].

It's important to note that the appearance of features in the FCC can vary depending on factors such as lighting conditions, atmospheric conditions, and the specific characteristics of the features themselves. Therefore, it's essential to consider these factors when interpreting the imagery and identifying features.

Additionally, you can use image processing techniques such as histogram equalization, contrast stretching, and spatial filtering to enhance the visibility of certain features in the FCC. These techniques can help you to better distinguish between different features and to extract accurate information from the image.

4.2 Image Resolution

Resolution refers to the level of detail that can be detected or displayed in an image. In remote sensing, there are three types of resolution: spatial, spectral, and temporal. The level of detail in an image, measured by the size of the pixels (picture elements) on the ground, also known as Ground Sample Distance (GSD) [11].

4.3 Spatial Resolution

Spatial resolution refers to the size of the smallest feature that can be detected by a satellite sensor or displayed in a satellite image [11]. It's usually presented as a single value representing the length of one side of a square. A smaller pixel size generally results in a more detailed image, but also leads to larger file sizes that can be difficult to process, store, and manage. For example, a spatial resolution of 250 m means that one pixel represents an area of 250 m by 250 m on the ground. Think of it like a puzzle. The smaller the puzzle pieces, the more detailed the picture will be. Similarly, the smaller the spatial resolution, the more detailed the image will be. Think of a pixel as a small square box that contains information about the area it represents. In this case, the box is 250×250 m, which is equivalent to an area of 62,500 square meters (or 250×250 m $= 62,500 m^2$). All the features within that 250×250 m area are represented by a single value. This value is usually a digital number that represents the brightness or reflectance of the area.

Imagine you're looking at a satellite image of a city. A single 250 m pixel might cover an entire city block, including buildings, roads, trees, and cars. But in the image, all those features are represented by a single value, which is the average brightness or reflectance of the entire area. This means that if you want to detect specific features within that 250×250 m area, such as individual buildings or trees, you might not be able to do so. The pixel is too large to capture those details.

For example, if you want to count the number of trees in a forest, a 250 m pixel might not be sufficient. You might need a higher resolution image, such as 10 or 1 m, to detect individual trees.

4.4 Temporal Resolution

Temporal resolution is crucial because it helps us understand how things change over time. Temporal resolution refers to the frequency at which remote sensing data is collected over a specific area. In other words, it's how often a satellite or sensor takes a picture of the same place. Think of it like taking a photo of a place every hour. The more frequently you take photos, the more you can see how things change over time. In remote sensing, temporal resolution works the same way. Satellites or sensors orbit the Earth, taking pictures of the same area at regular intervals. The frequency of these intervals determines the temporal resolution [11].

- **High Temporal Resolution**: A satellite takes pictures of the same area every hour, day, or week. This is useful for monitoring changes that happen quickly, like weather patterns or crop growth.
- **Low Temporal Resolution**: A satellite takes pictures of the same area every month, year, or decade. This is useful for monitoring changes that happen slowly, like land use changes or glacier movement.

By analysing data collected at different times, we can:

- Monitor crop health and growth
- Track changes in weather patterns and climate
- Study the movement of glaciers and sea ice
- Detect natural disasters like floods and landslides
- Forecast the weather events

4.5 Spectral Resolution

Spectral resolution refers to the ability of a satellite sensor to measure specific wavelengths of the electromagnetic spectrum. The finer the spectral resolution, the narrower the wavelength range for a particular channel or band [11].

Imagine you're looking at a rainbow. A sensor with high spectral resolution can detect the individual colours of the rainbow, while a sensor with low spectral resolution might only detect the overall brightness of the rainbow. In addition, if we consider the Landsat series, comprising Landsat 5 (launched in 1984), Landsat 7 (launched in 1999), and Landsat 8 (launched in 2013), we can see that Landsat 5 can detect 7 bands, Landsat 7 detects 8 different bands (adding a new band for detecting thermal infrared radiation, or heat), and Landsat 8 detects 11 different wavelengths of light (In addition to these 7 bands, Landsat 8 also features a panchromatic band, which captures black-and-white imagery, and a cirrus cloud band, designed to detect high-level clouds. Furthermore, the satellite's Thermal Infrared Sensor (TIRS) collects data in two thermal infrared bands. To extract meaningful information from these bands, researchers often combine them in various ways). Among these three, Landsat 8 has the highest spectral resolution, followed by Landsat 7, and then Landsat 5. This means that Landsat 8 can provide more detailed and accurate information about the environment than its predecessors [14]. Landsat 9 was also launched on September 27, 2021. Landsat 9, like its predecessor Landsat 8, offers a high spectral resolution with 11 bands, enabling detailed Earth observations. However, Landsat 9 features improved radiometric performance, providing more precise data for monitoring subtle changes in land and water surfaces.

4.6 Vector and Raster Types

The format of the data, with vector data consisting of points, lines, and polygons, and raster data composed of a grid of pixels. Understanding the differences between these data types is essential for effective data management, analysis, and visualization.

Vector data: Vector data represents geographic features as a collection of points, lines, and polygons. These features are defined by a set of x, y coordinates, which are stored as a series of vertices. Vector data is often used to represent discrete objects, such as boundaries (e.g., country borders, property lines), roads and highways, buildings and infrastructure and any points of interest (e.g., landmarks, amenities). Vector data is typically stored in a format such as shapefiles (.shp), GeoJSON (.geojson), .kmz etc.

Vector data offers several advantages, including high precision, scalability, and efficient storage. It can represent features with exact coordinates, making it ideal for applications requiring precise spatial accuracy, and remains crisp and clear at any zoom level without sacrificing detail or clarity. Additionally, vector data requires less storage space compared to raster data, especially for simple features, making it a storage-efficient option.

Raster data: Raster data represents geographic features as a grid of pixels, each with a specific value or attribute, and is often used to represent continuous phenomena such as imagery, elevation and terrain models, climate and weather data, and land cover and land use classification, typically stored in formats like GeoTIFF, and ASCII Grid. The advantages of raster data include its efficient representation of large, continuous datasets, making it ideal for applications like image processing and terrain analysis, its simple grid-based structure that makes it easy to process and analyse, and its wide applicability in various fields, ranging from environmental monitoring to urban planning.

Raster data does have some limitations. These include limited precision due to the resolution of the grid, which can result in a loss of precision and accuracy. Additionally, there are large storage requirements, particularly for high-resolution datasets. This can be a significant drawback, especially when working with large-scale projects or datasets.

The vector and raster representations of point, line, and polygon features are illustrated in Figs. 6, 7 and 8. A notable difference between the two representations can be observed in the level of detailing, particularly in the sharpness of the edges. In the vector representation, the edges appear crisp and well-defined, whereas in the raster representation, the edges appear more pixelated and less refined. This distinction highlights the fundamental differences between vector and raster data models, with vector data capable of capturing precise geometric details and raster data relying on a grid-based approximation (Figs. 6, 7 and 8).

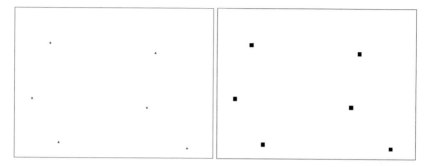

Fig. 6 Representation of objects using point feature (vector and raster)

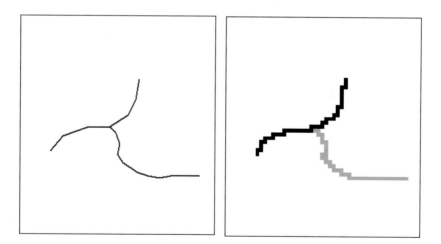

Fig. 7 Representation of objects using line feature (vector and raster)

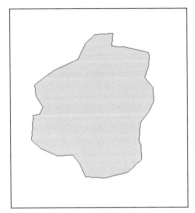 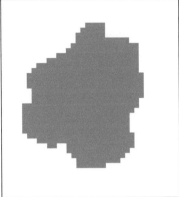

Fig. 8 Representation of area using polygon feature (vector and raster)

5 Various Kinds of Remote Sensing Data

Remote sensing data comes in different forms, each with its unique characteristics, advantages, and limitations. These variations in remote sensing data are critical in determining their suitability for specific projects and applications.

5.1 Digital Elevation Models (DEMs)

These are 3D representations of the Earth's surface, providing information on topography, elevation, and relief. DEMs are essential for terrain analysis such as morphometric study, hydrological modelling, and infrastructure planning. Here are some popular DEM data sources:

Digital Elevation Model (DEM) Data Sources

a. **SRTM (*Shuttle Radar Topography Mission*)**: Provided by NASA, SRTM DEM data has a resolution of 30 m and covers most of the world with an absolute vertical height accuracy of less than 16 m. You can select your area of interest and download the data in various formats.

You can download SRTM DEM data freely from the following sources such as USGS Earth Explorer, Geoprocessing online, and Open Topography.

- **USGS Earth Explorer**: This is a popular source for SRTM DEM data, with a resolution of 30 m. You can select your area of interest and download the data. Additionally,

the data is provided in GeoTIFF format, which is a common format for geospatial raster data.
- **Geoprocessing Online**: This website allows you to download high-resolution SRTM DEM data with a resolution of 1-arc second (30 m). You can define your area of interest by uploading a vector geodata file or by drawing the area on a map.
- **Open Topography**: This website provides SRTM DEM data with varying resolutions, including 30 and 90 m. You can search for your area of interest and download the data in various formats.

b. *ASTER Global Digital Elevation Model*: Offered by NASA, ASTER GDEM has a global resolution of 90 m and a resolution of 30 m in the United States. It can be downloaded from NASA Earthdata or the USGS Earth Explorer.

c. *JAXA's Global ALOS 3D World*: It also known as the ALOS Global Digital Surface Model (AW3D30), is a global dataset generated from images collected using the Panchromatic Remote-Sensing Instrument for Stereo Mapping (PRISM) aboard the Advanced Land Observing Satellite (ALOS) from 2006 to 2011 [15]. The Japan Aerospace Exploration Agency (JAXA) releases this global digital surface model (DSM) dataset with a horizontal resolution of approximately 30-m mesh (1 arcsec) free of charge.

The dataset has been compiled with images acquired by the Advanced Land Observing Satellite "DAICHI" (ALOS). The dataset is published based on the DSM dataset (5-m mesh version) of the "World 3D Topographic Data", which is the most precise global-scale elevation data at this time, and its elevation precision is also at a world-leading level as a 30-m mesh version.

This dataset is expected to be useful for scientific research, education, as well as the private service sector that uses geospatial information. The dataset is available in two versions, average and median, created from the original 5-m mesh using different downsampling methods. However, only the average version is available from OpenTopography.

The latest version of AW3D30 is 3.2, released in January 2021, which can be downloaded from the JAXA website. The dataset has undergone several updates, with the initial version 1 released in May 2016, covering the global terrestrial region (within approximately 82 deg. of N/S latitudes) with approximately 22,100 tiles.

d. *ICESat/GLAS*: Provided by NASA, ICESat/GLAS DEM data has a resolution of 30 m and covers the entire globe.
e. *MOLA DEM of Mars*: Provided by the USGS Astrogeology Science Center, MOLA DEM data has a resolution of 463 m and covers the entire planet of Mars.
f. *Cartosat DEM*: Cartosat DEM refers to the elevation data generated by India's Cartosat series of Earth observation satellites. The Cartosat satellites, launched by the Indian

Space Research Organisation (ISRO), are equipped with high-resolution panchromatic cameras capable of capturing detailed images of the Earth's surface. Bhuvan website allows you to download the Cartosat DEM data of 30 m resolution freely.

These are just a few examples of the many DEM data sources available. Each has its own strengths and weaknesses, and the choice of which one to use depends on the specific application and requirements.

5.2 Digital Terrain Models (DTMs)

Digital Terrain Models (DTMs) are computer-generated representations of the Earth's surface, created by combining numerous elevation data points. These models provide a precise and detailed depiction of the landscape, allowing users to analyze and understand the terrain's complexities. DTMs are constructed by interpolating elevation values from a set of discrete points, which can be obtained through various methods such as LiDAR surveys, photogrammetry, or other remote sensing technologies. The resulting model is a digital dataset that can be used for a range of applications, including terrain analysis, hydrological modeling, environmental monitoring, and urban planning. There are different types of DTMs, including Digital Elevation Models (DEMs) that represent the bare earth surface and Digital Surface Models (DSMs) that include features such as buildings and vegetation. The accuracy and resolution of a DTM depend on the quality and density of the input data, as well as the interpolation method used. By providing a detailed and accurate representation of the terrain, DTMs have become an essential tool for various fields, enabling users to make informed decisions and gain valuable insights [16].

5.3 Multispectral Aerial Imageries

Multispectral imageries are a type of remote sensing data that capture images of the Earth's surface in multiple spectral bands, providing valuable information on land cover, vegetation health, and environmental conditions. These images are acquired by sensors mounted on satellites, aircraft, or drones, which detect reflected radiation in specific wavelengths of the electromagnetic spectrum. The resulting data is used in a wide range of applications, including crop monitoring, land use classification, and natural resource management.

One of the primary advantages of multispectral imageries is their ability to distinguish between different land cover types, such as forests, grasslands, and urban areas. For example, the Normalized Difference Vegetation Index (NDVI) is a widely used metric that calculates the difference between the near-infrared and red spectral bands to estimate

5 Various Kinds of Remote Sensing Data

vegetation health and density. This information is essential for monitoring crop growth, detecting deforestation, and tracking changes in land use patterns.

For instance, in agriculture, multispectral imageries can be used to estimate crop yields, detect crop stress, and identify areas that require irrigation or fertilization. This information enables farmers to make informed decisions, reducing costs and increasing productivity. In land use classification, multispectral imageries can be used to identify different land cover types, such as forests, grasslands, or urban areas. This information is essential for urban planning, conservation efforts, and natural resource management. For example, in natural resource management, multispectral imageries can be used to monitor deforestation, track changes in land cover, and identify areas that require conservation efforts. This information enables policymakers to make informed decisions, reducing the impact of human activities on the environment.

Examples of multispectral remote sensing:

1. **Landsat 8**: Landsat 8 is a NASA satellite that captures multispectral images of the Earth's surface in eight different bands, including visible, near-infrared, and thermal infrared. The data is used for land use classification, crop monitoring, and natural resource management.
2. **WorldView-4**: WorldView-4 is a commercial satellite operated by DigitalGlobe that captures multispectral images in eight bands, including visible, near-infrared, and shortwave infrared. The data is used for applications such as crop monitoring, land use classification, and environmental monitoring.
3. **Sentinel-2**: Sentinel-2 is a European Space Agency (ESA) satellite that captures multispectral images in 13 bands, including visible, near-infrared, and shortwave infrared. The data is used for land cover classification, crop monitoring, and natural disaster management.
4. **MODIS**: MODIS (Moderate Resolution Imaging Spectroradiometer) is a NASA instrument on board the Terra and Aqua satellites that captures multispectral images in 36 bands, including visible, near-infrared, and thermal infrared. The data is used for applications such as climate modeling, vegetation health monitoring, and ocean colour monitoring.
5. **ASTER**: ASTER (Advanced Spaceborne Thermal Emission and Reflection Radiometer) is a NASA instrument aboard the Terra satellite that captures multispectral images in 14 bands, including visible, near-infrared, and thermal infrared. The data are used for applications such as land surface temperature monitoring, vegetation health monitoring, and geological mapping.
6. **Planet Labs**: Planet Labs is a commercial satellite constellation that captures multispectral images in four bands (blue, green, red, and near-infrared) with a resolution of 3–5 m. The data is used for applications such as crop monitoring, land use classification, and environmental monitoring.

7. **RapidEye**: RapidEye is a commercial satellite constellation that captures multispectral images in five bands (blue, green, red, red-edge, and near-infrared) with a resolution of 5 m. The data is used for applications such as crop monitoring, land use classification, and environmental monitoring.
8. **GeoEye-1**: GeoEye-1 is a commercial satellite operated by DigitalGlobe that captures multispectral images in eight bands, including visible, near-infrared, and shortwave infrared. The data is used for applications such as crop monitoring, land use classification, and environmental monitoring.
9. **IKONOS**: IKONOS is a commercial satellite operated by DigitalGlobe that captures multispectral images in four bands (blue, green, red, and near-infrared) with a resolution of 1 m. The data is used for applications such as land use classification, crop monitoring, and environmental monitoring.
10. **UAV-based multispectral imaging**: Unmanned Aerial Vehicles (UAVs) equipped with multispectral cameras can capture high-resolution images of crops, forests, and other areas of interest. The data is used for applications such as crop monitoring, precision agriculture, and environmental monitoring.

5.4 Administrative Boundary Maps

These are spatial datasets that define the boundaries of administrative units such as countries, states, cities, and districts. These maps are created by combining remote sensing data with geographic information systems (GIS) and other spatial data to create a comprehensive and accurate representation of a region's administrative boundaries. They are essential for governance, urban planning, and socio-economic analysis.

There are several remote sensing data sources for administrative boundary maps, including:

- **DIVA-GIS**: A free computer program for mapping and geographic data analysis that provides free spatial data for the whole world, including administrative boundaries, inland water, roads, and railroads. Downloads are available by country.
- **Natural Earth**: A source of excellent base data created by cartographers for cartographers, with worldwide coverage at a variety of scales.
- **OpenStreetMap**: A great source for street data and other infrastructure, with global coverage. Data can be downloaded directly through the map interface or other sources.
- **GADM Database of Global Administrative Areas**: A database that provides GIS files for the world's administrative boundaries, including countries and lower-level subdivisions.
- **ArcGIS Living Atlas of the World**: A collection of geographic information from around the globe with a wide range of topics.

5.5 Meteorological Data

With the advent of remote sensing technology, meteorological data is now available in spatial formats, enabling the analysis of weather patterns, climate trends, and environmental phenomena. Sources like TRMM (Tropical Rainfall Measuring Mission) provide precipitation data, while others like MODIS (Moderate Resolution Imaging Spectroradiometer) offer atmospheric and oceanic data. TRMM's satellite-based sensors collect data on rainfall rates, storm intensity, and other precipitation-related metrics, enabling researchers to better understand global weather patterns and climate trends. MODIS sensors aboard NASA's Terra and Aqua satellites collect data on variables such as sea surface temperature, atmospheric aerosols, and ocean currents, providing insights into the complex interactions between the atmosphere, oceans, and land surfaces.

Apart from these there are different climate project models that utilize remote sensing data:

VITO Remote Sensing Climate Project Model: This model uses historical time series of satellite data to analyse changes over time and study the effects of climate change on ecosystems. It also monitors the effectiveness of adaptation measures implemented to mitigate the impacts of climate change.

Remote Sensing Systems (RSS) Climate Project Model: This model uses satellite data records to evaluate multi-decadal changes in the climate system. It intercalibrates measurements from different satellites to produce a single, long-term data record. The model has been used to study atmospheric temperature, total column water vapor, and wind speed.

Upper Air Temperature Measurement Climate Project Model: This model uses tropospheric temperature datasets, such as TLT (Temperature Lower Troposphere), TMT (Temperature Middle Troposphere), and TTT (Temperature Total Troposphere), to investigate changes in the tropospheric temperature over the last 35 years. It compares the results with the predictions of climate models to validate the models' accuracy.

These climate project models demonstrate the power of remote sensing data in understanding and predicting climate change. By utilizing these models, researchers can better analyze climate trends, identify drivers of climate change, and inform decision-making processes related to environmental management and policy.

6 Geographic Information Systems (GIS)

Geographic Information System (GIS) is computer-based tool that facilitates the collection, storage, interpretation, analysis, and visualization of geographical (geographically referenced data, i.e., data with location information) data. GIS integrates various data types, including satellite imagery, topographic maps, and demographic statistics, into a clear framework.

GIS software is essential for handling geospatial data because, unlike image editing software such as AutoCAD, Paint, and Photoshop, it is specifically designed to analyse and manipulate geospatial data, taking into account the unique characteristics of spatial relationships and geographic coordinates. While image editing software can edit and enhance visual aspects of images, GIS software provides advanced tools for spatial analysis, mapping, and modelling, enabling users to extract meaningful insights and patterns from geospatial data [1].

Key Features of GIS Software:

- **Spatial Analysis**: Perform advanced spatial analysis, including spatial joins, buffers, and network analysis.
- **Data Integration**: Integrate multiple data sources, including satellite imagery, GPS data, and tabular data.
- **Mapping**: Create interactive maps and visualize geospatial data.
- **3D Analysis**: Perform 3D analysis and visualization of geospatial data.
- **Programming**: Offer programming interfaces, such as Python and R, for automating tasks and developing custom applications.

Here are some popular GIS software, categorized into proprietary (paid softwares) and non-proprietary (Open access)

Proprietary Software:

1. **ERDAS Imagine**
2. **ENVI**
3. **PCI Geomatics**
4. **ArcGIS**: A GIS software package developed by ESRI offering advanced spatial analysis and mapping capabilities.
5. **ILWIS**
6. **TerrSet**

Non-Proprietary Software:

1. **QGIS**
2. **GRASS GIS**:
3. **Orfeo ToolBox**
4. **Google Earth Engine**: A cloud-based platform for remote sensing and GIS analysis
5. **DIVA GIS**

7 Integration of GIS and Remote Sensing

The integration of GIS and Remote Sensing enhances the capabilities of both technologies. Satellite imagery provides up-to-date and large-scale observations, while GIS offers the tools to analyse these observations within a spatial context. This synergy is crucial for monitoring changes in water resources over time and across vast regions [2].

The contemporary world faces numerous water-related challenges, including climate change, population growth, urbanization, and pollution. GIS and Remote Sensing offer critical solutions to these challenges. They provide data-driven insights that support adaptive management strategies, ensuring resilience and sustainability in water resource management.

Climate change, for instance, alters precipitation patterns and increases the frequency of extreme weather events, impacting water availability and quality. Remote Sensing can track these changes in real-time, while GIS models can predict their long-term effects. Similarly, in urban areas, GIS helps manage stormwater, reduce flooding risks, and optimize the distribution of water infrastructure [2].

Key Uses:

- Map watersheds, rivers, streams, and aquifers with high accuracy.
- Water resource management: Remote Sensing and GIS can be used to monitor water quality, track changes in water bodies, and optimize water distribution systems. In addition to develop adaptive management strategies for water resource management and conservation.
- Flood risk assessment: Remote Sensing and GIS can be used to identify flood-prone areas, model flood scenarios, and develop early warning systems.
- Climate Change Impact Assessment: Remote Sensing and GIS can be used to track changes in climate patterns, model the impacts of climate change on water resources, and develop adaptation strategies.
- Environmental Monitoring: Remote Sensing and GIS can be used to monitor air and water quality, track changes in ecosystems, and develop early warning systems for environmental hazards.
- Sustainable Urban Planning: Remote Sensing and GIS can be used to optimize urban planning, reduce the urban heat island effect, and develop sustainable transportation systems.
- Disaster Response and Recovery: Remote Sensing and GIS can be used to assess damage, prioritize response efforts, and develop recovery plans after natural disasters.
- Ecosystem Management: Remote Sensing and GIS can be used to monitor ecosystem health, track changes in biodiversity, and develop conservation strategies.
- Agricultural Management: Remote Sensing and GIS can be used to optimize crop yields, monitor soil health, and develop precision agriculture strategies.

- Coastal Zone Management: Remote Sensing and GIS can be used to monitor coastal erosion, track changes in sea level, and develop sustainable coastal zone management plans.
- Assess erosion and runoff in various watershed scales.
- Evaluate the impacts of land-use changes on water resources.
- Identify landslide-prone areas and drought zones.

8 Conclusion

This chapter has provided a detailed overview of the fundamentals of remote sensing and GIS, highlighting their unique characteristics and capabilities. The discussion covered various remote sensing platforms and data sources, including satellite imagery, aerial photography, and ground-based observations, each offering distinct advantages for different applications. The integration of remote sensing and GIS has proven to be a powerful tool for water resources management, enabling the identification of sustainable water sources, assessment of environmental impact, and development of conservation strategies.

References

1. Chowdary, V. M., Ramakrishnan, D., Srivastava, Y. K., Chandran, V., & Jeyaram, A. (2009). Integrated water resource development plan for sustainable management of Mayurakshi watershed, India using remote sensing and GIS. *Water resources management, 23*(8), 1581–1602.
2. Masud, M. J., & Bastiaanssen, W. G. (2017). Remote sensing and GIS applications in water resources management. *Water resources management*, 351–373.
3. Dwivedi, R. S., K. Sreenivas, K. V. Ramana, P. R. Reddy, and G. Ravi Sankar. "Sustainable development of land and water resources using geographic information system and remote sensing." *Journal of the Indian Society of Remote Sensing* 34 (2006): 351–367.
4. Chen, Y., Takara, K., Cluckie, I. D., & De Smedt, F. H. (2004). GIS and remote sensing in hydrology. *Water Resources and Environment*.
5. Elachi, C., & Van Zyl, J. J. (2021). *Introduction to the physics and techniques of remote sensing*. John Wiley & Sons.
6. Lapaine, M., & Frančula, N. (2001). Cartography and Remote Sensing. *Bilten Znanstvenog vijeća za daljinska istraživanja i fotointerpretaciju HAZU, 15*(16), 145–154.
7. Jia, J., Sun, H., Jiang, C., Karila, K., Karjalainen, M., Ahokas, E., ... & Hyyppä, J. (2021). Review on active and passive remote sensing techniques for road extraction. *Remote Sensing, 13*(21), 4235.
8. Zhu, L., Suomalainen, J., Liu, J., Hyyppä, J., Kaartinen, H., & Haggren, H. (2018). A review: Remote sensing sensors. *Multi-purposeful application of geospatial data, 19*.
9. Prasad, S., Bruce, L. M., & Chanussot, J. (2011). Optical remote sensing. *Advances in Signal Processing and Exploitation Techniques*.
10. Deng, L., Mao, Z., Li, X., Hu, Z., Duan, F., & Yan, Y. (2018). UAV-based multispectral remote sensing for precision agriculture: A comparison between different cameras. *ISPRS journal of photogrammetry and remote sensing, 146*, 124–136.

References

11. Held, A., Ticehurst, C., Lymburner, L., & Williams, N. (2003). High resolution mapping of tropical mangrove ecosystems using hyperspectral and radar remote sensing. *International Journal of Remote Sensing, 24*(13), 2739–2759.
12. Aggarwal, S. (2004). Principles of remote sensing. *Satellite remote sensing and GIS applications in agricultural meteorology, 23*(2), 23–28.
13. Bajwa, R.S., Ahsan, N. & Ahmad, S.R. A Review of Landsat False Color Composite Images for Lithological Mapping of Pre-Cambrian to Recent Rocks: A Case Study of Pail/Padhrar Area in Punjab Province, Pakistan. *J Indian Soc Remote Sens* **48**, 721–728 (2020). https://doi.org/10.1007/s12524-019-01090-7.
14. Wulder, M. A., Loveland, T. R., Roy, D. P., Crawford, C. J., Masek, J. G., Woodcock, C. E., ... & Zhu, Z. (2019). Current status of Landsat program, science, and applications. *Remote sensing of environment, 225*, 127–147.
15. Tadono, T., Takaku, J., Tsutsui, K., Oda, F., & Nagai, H. (2015, July). Status of "ALOS World 3D (AW3D)" global DSM generation. In *2015 IEEE international geoscience and remote sensing symposium (IGARSS)* (pp. 3822–3825). IEEE.
16. Galin, E., Guérin, E., Peytavie, A., Cordonnier, G., Cani, M. P., Benes, B., & Gain, J. (2019, May). A review of digital terrain modeling. In *Computer Graphics Forum* (Vol. 38, No. 2, pp. 553–577).

Land Use Land Cover Change Detection Using Google Earth Engine

Abstract

Land Use and Land Cover (LULC) change detection is a critical component in understanding the dynamics of landscape transformations and their impact on environmental sustainability. This chapter provides the history and methodologies associated with LULC change detection using Remote Sensing (RS), Geographic Information Systems (GIS), and Machine Learning (ML) techniques. It covers the evolution of LULC change detection methods, from traditional techniques to the integration of ML algorithms in recent years. The chapter delves into the principles and applications of both supervised and unsupervised machine learning models for LULC classification, highlighting their strengths and limitations. A case study on the Upper Bhavani Basin for the period 2017–2021 demonstrates the practical application of these methods, utilizing Landsat 8 images and the Random Forest classifier in Google Earth Engine (GEE). The study assesses the accuracies and Kappa coefficient values to evaluate the performance of the model.

Keywords

Google Earth Engine • Random forest • Kappa coefficient • Remote sensing

1 Introduction

Land use land cover (LULC) change detection is a crucial aspect of environmental monitoring, urban planning, and natural resource management. Accurate detection and analysis of changes in LULC patterns over time are crucial for understanding the environmental impacts of human activities, predicting the likelihood of natural disasters, and developing sustainable development strategies [1].

Remote sensing (RS), Geographic Information Systems (GIS), and machine learning (ML) are powerful tools that have transformed the field of LULC change detection. Remote sensing provides a cost-effective and efficient means of collecting data over large areas, while GIS enables the integration and analysis of spatial data. Machine learning algorithms, on the other hand, offer a robust approach to classifying and predicting LULC changes. The integration of these technologies has enabled researchers and practitioners to detect LULC changes with high accuracy and to identify the underlying drivers of these changes [2].

2 History of LULC Change Detection with RS, GIS, and ML

In early beginnings, when remote sensing emerged (during 1960s), the LULC change detection was performed through visual interpretation of aerial satellite imageries. This was a labour-intensive process that relied on human analysts to manually examine the images and identify changes. It was a crucial step in the development of LULC change detection, but it had its limitations, such as being time-consuming and prone to human errors [1, 2].

However, with the widespread availability of GIS software and technology, the process became significantly more efficient. This enabled the storage, analysis, and interpretation of geospatial data in a shorter time, making LULC change detection easier and more accurate [2].

Later in 2000s, advancement in remote sensing technology led to the availability of high-resolution satellite imageries, which helped the researchers to detect the LULC changes with greater accuracy [3].

In 2010s, Google Earth Engine (GEE) was launched, providing a cloud-based platform for remote sensing data analysis. This enabled researchers to access and process large datasets, and integrate machine learning (ML) algorithms to analyse LULC changes [4, 5].

The integration of GEE platforms in LULC change detection has transformed the field by offering several advantages over traditional GIS-based approaches. While sophisticated GIS software has been widely used in LULC change detection, Google Earth Engine (GEE) offers several advantages that make it an attractive alternative. One of the primary benefits of GEE is its ability to handle large datasets and perform complex analyses at a scale that is not feasible with traditional GIS software. This is particularly important in LULC change detection, where large areas need to be analysed over extended periods.

Another significant advantage of GEE is its cloud-based infrastructure, which eliminates the need for expensive hardware and software investments. This makes it an ideal choice for researchers and organizations with limited resources. Additionally, GEE provides access to a vast repository of satellite imagery and other geospatial data, which can be easily integrated into LULC change detection workflows. Above all, the algorithms used to perform LULC change analysis can be continuously improved and updated as new data becomes available, ensuring that they remain accurate and effective over time [5].

GEE's simplicity and user-friendly interface also make it more accessible to researchers who may not have programming or GIS expertise. This is in contrast to sophisticated GIS software, which often requires advanced technical skills to operate effectively. Furthermore, GEE's ability to perform tasks in parallel and its integration with other Google services, such as Google Cloud Storage, Earth Engine data catalogues and Google Cloud Computing, make it an ideal platform for large-scale LULC change detection projects. This enables researchers to process large datasets quickly and efficiently, reducing the time and effort required for analysis.

For example, a study published in Remote Sensing [6] used GIS, remote sensing, and machine learning to highlight the correlation between LULC changes and flash-flood potential. The study demonstrated the effectiveness of using machine learning algorithms to classify LULC changes and to predict flash-flood potential. Similarly, another study [7] reviewed the application of machine learning algorithms in LULC change detection and highlighted their potential in improving the accuracy of LULC change detection. Overall, the integration of remote sensing, GIS, and machine learning has greatly enhanced our ability to detect and analyse LULC changes, and proved as a better option for the development of more effective strategies for sustainable development.

3 Machine Learning in LULC Classification

Google Earth Engine (GEE) is a powerful platform that provides a vast amount of remote sensing data and tools for LULC classification. In this context, Machine Learning (ML) algorithms play a vital role in improving the accuracy and efficiency of LULC classification [7]. There are two main types of ML algorithms used for LULC classification: supervised and unsupervised algorithms.

1. Supervised ML algorithm
2. Unsupervised ML algorithm

3.1 Supervised ML Algorithms

Supervised ML algorithms are trained on labelled data, where the target class labels are known. Supervised machine learning for LULC classification is a type of machine learning where the algorithm is trained on a labelled dataset, known as the **ground truth data**, which consists of remote sensing or Earth observation data (e.g., satellite or aerial imagery) and corresponding **LULC class labels**. The ground truth data serves as a reference point for the algorithm to learn from, allowing it to identify patterns and relationships between the remote sensing data and the LULC class labels. The goal of supervised machine learning for LULC classification is to train a model that can accurately predict the LULC class labels for new, unseen remote sensing data [8, 9].

In supervised machine learning, the ground truth data is used to:

- Train the model to recognize patterns and relationships between the input data and output labels
- Evaluate the performance of the model using metrics such as accuracy, precision, and recall
- Refine and update the model over time to maintain its accuracy and performance

The LULC class labels in the ground truth data typically include categories such as:

- Urban areas
- Agricultural land
- Forests
- Grasslands
- Water bodies

Some popular supervised ML algorithms for LULC classification using GEE include:

A. **Random Forest**

Random Forest is a popular machine learning algorithm used for classification and regression tasks. It's a collective learning method that combines multiple decision trees to produce a more accurate prediction model [8, 9]. In the context of remote sensing, Random Forest is often used for LULC classification, where the goal is to assign a class label (e.g., water, vegetation, urban) to each pixel in an image. The detailed explanation of the Random Forest classifier in LULC classification, with examples in each step, using Google Earth Engine (GEE) is given in this section.

RF is the collaborative learning method that combines multiple decision trees to improve the accuracy of the classification model. This Algorithm comprises

1. *Preparation of Datasets*

This is to prepare the remote sensing data for classification by splitting it into training and testing datasets. For example; if we want to classify waterbodies in a region using Landsat 8 imagery. We can use the GEE platform to collect a time series of Landsat 8 images for the region, and then split the data into training and testing datasets. For instance, we can use 80% of the data for training and 20% for testing. When we split the data, we're creating two separate datasets: one for training the model and another for evaluating its performance.

2. *Feature Extraction*

To extract relevant features from the remote sensing data that can help distinguish between different LULC classes. For example, in the waterbody classification, we can extract the features such as:

- *Spectral bands*: using the Blue, Green, Red, and NIR bands to distinguish between water and land pixels.
- *Texture features*: calculating the mean, variance, and homogeneity of the spectral bands to capture the spatial patterns of waterbodies.
- *Spatial features*: calculating the NDVI (Normalized Difference Vegetation Index) to distinguish between water and vegetation pixels.

The raw remote sensing data is transformed into a format that's more suitable for classification when features are extracted. A set of numerical values that describe the characteristics of each pixel is essentially created, which can then be used to train a classification model.

3. *Decision Tree Construction*

In this step, we construct multiple decision trees using the training data, with each tree randomly selecting a subset of features and samples. For example, in the waterbody classification as mentioned above, we can construct multiple decision trees using the training data, with each tree randomly selecting a subset of the spectral, texture, and spatial features. For instance, one tree might use the Blue and NIR bands, while another tree might use the texture features and NDVI.

4. *Classification*

The predictions of multiple decision trees are combined to produce a final classification. For example, in the waterbody classification, with the Random Forest algorithm to combine the predictions of multiple decision trees. The predictions of multiple trees are combined through a voting mechanism, where each pixel's class label is voted on by the multiple trees, and the tree with the most votes is determined, resulting in the pixel being assigned the corresponding class label. This process helps to reduce the noise and variability in the individual tree predictions, producing a more accurate classification. For instance, if 80% of the trees predict a pixel as water, and 20% predict it as land, the final classification would be water.

B. **Support Vector Machines (SVM)**

Support Vector Machine, which is a popular machine learning algorithm used for classification and regression analysis. SVM is a type of supervised learning algorithm that aims to find a decision boundary or hyperplane that separates the data into different classes.

In the context of LULC classification, SVM is used to classify pixels in satellite imagery data into different land use and land cover classes, such as water, urban, forest, and agricultural land. SVM works by finding the optimal hyperplane that maximizes the margin between the classes, thereby minimizing the error [10].

The key advantages of SVM include:

- **Robustness to noise and outliers**: SVM is robust to noisy data and outliers, making it suitable for real-world applications.
- **Flexibility**: SVM can be used for both linear and non-linear classification problems.
- **High accuracy**: SVM has been shown to achieve high accuracy in various classification tasks.

Steps involved in SVM methodology for LULC classification using GEE

- Data acquisition and preprocessing
- Training data preparation
- SVM classifier training
- LULC classification
- Accuracy assessment

In general, the SVM methodology can be applied to any study area to classify land use and land cover into different classes. The accuracy of the classification can be assessed by comparing the classified map with the ground truth data.

a. **Data Acquisition and Preprocessing**

The first step is to acquire satellite imagery data and preprocess it to prepare it for classification. This may involve tasks such as data cleaning, normalization, and feature extraction.

b. **Training Data Preparation**

This involves collecting ground truth data, which is used to train the SVM classifier. The training data should be representative of the different land use and land cover classes present in the study area.

c. SVM Classifier Training

The SVM classifier is then trained using the training data. This involves selecting the appropriate kernel function and tuning the hyperparameters to optimize the performance of the classifier.

d. LULC Classification

Once the classifier is trained, it can be applied to the pre-processed satellite imagery data to perform the LULC classification. This involves using the trained classifier to predict the land use and land cover classes for each pixel in the image.

e. Accuracy Assessment

Finally, the accuracy of the classification is assessed using metrics such as overall accuracy, producer's accuracy, and user's accuracy. This involves comparing the classified image with the ground truth data to evaluate the performance of the classifier.

C. Minimum Distance Classifier

It is a simple and widely used classification algorithm in remote sensing and image processing. The Minimum Distance Classifier is a **supervised** classification method, which means that it requires labelled training data to learn the patterns and relationships between the features and classes. The algorithm uses the labelled training data to calculate the mean vector of each class and then classifies the unknown pixels based on the minimum distance to the class mean [11, 12].

In the context of LULC classification, the Minimum Distance Classifier is used to classify pixels in satellite imagery data into different land use and land cover classes, such as water, urban, forest, and agricultural land. The algorithm works by calculating the distance between the unknown pixel and the mean vector of each class in the feature space. This methodology is very simple to implement and faster in performance, but it is not suitable for complex non-linear datasets [12].

Steps involved in Minimum Distance Classifier for LULC classification using GEE
Step 1: Data Acquisition and Preprocessing

- Collect satellite imagery data for the study area using GEE's data catalog or upload your own data.
- Preprocess the data by converting it to a suitable format, such as GeoTIFF, and performing radiometric and atmospheric corrections (e.g., cloud correction).

Step 2: Training Data Preparation
In this step, we collect and prepare labelled training data for each LULC class. This involves:

- Collecting a set of known pixels for each LULC class (e.g., water, urban, forest, agricultural land)
- Extracting relevant spectral features from the training data (e.g., spectral bands, vegetation indices)
- Creating a feature space that represents the characteristics of each LULC class

Step 3: Calculate Class Mean Vectors
The goal of this step is to create a set of mean vectors that can be used to represent each LULC class in the feature space.

Step 4: Minimum Distance Classification
In this step, we perform the minimum distance classification by calculating the distance between each unknown pixel and the mean vector of each class in the feature space. This involves:

- Calculating the distance between each pixel and the mean vector of each class using a distance metric (e.g., Euclidean distance)
- Assigning the pixel to the class with the minimum distance

Step 5: Accuracy Assessment
The accuracy of the classification is assessed using metrics such as overall accuracy, producer's accuracy, and user's accuracy. This involves comparing the classified image with the ground truth data to evaluate the performance of the classifier.

D. CART (Classification and Regression Trees)

A decision tree-based algorithm that is simple to implement and interpret [13].
    The steps involved in the CART (Classification and Regression Trees) methodology for LULC classification [13], with examples:

Step 1: Data Preparation
In this step, we prepare the data for CART analysis by collecting and preprocessing the satellite imagery data and the corresponding LULC class labels. This involves:

- Importing the satellite imagery data into a GEE platform
- Preprocessing the data to remove clouds and errors

3 Machine Learning in LULC Classification

- Collecting and preparing the LULC class labels for each pixel, such as:
  - Waterbodies (e.g., lakes, rivers, oceans)
  - Urban areas (e.g., cities, towns, buildings)
  - Forests
  - Agricultural land

Step 2: Decision Tree Construction

In this step, we construct a decision tree by recursively partitioning the data into smaller subsets based on the values of the predictor variables. This involves:

- Selecting the best predictor variable to split the data at each node, such as:
  - Spectral band values (e.g., red, green, blue)
  - Vegetation indices (e.g., NDVI, EVI, NDWI)
  - Texture features
- Creating a decision rule based on the selected variable, such as:
  - "If NDVI > 0.5, then go to node 1" (higher NDVI values indicate more vegetation)
- Partitioning the data into smaller subsets based on the decision rule, such as:
  - Node 2: Vegetated areas (e.g., forests, agricultural land)

Step 3: Node Splitting

The goal of this step is to create a hierarchical structure of nodes that can be used to classify new pixels into different LULC classes. In this step, we split each node into two child nodes based on the values of the predictor variables. This involves:

- Selecting the best predictor variable to split the node, such as:
  - Spectral band values in the near-infrared range
  - Texture features
- Creating a decision rule based on the selected variable
- Splitting the node (Say Node 1) into two child nodes based on the decision rule, such as:
  - Node 4: Forests with high vegetation density
  - Node 5: Agricultural land with low vegetation density

Step 4: Leaf Node Assignment

The goal of this step is to assign each leaf node to a specific LULC class based on the patterns and relationships learned from the training data. In this step, we assign each leaf node to a specific LULC class based on the majority vote of the training pixels. This involves:

- Counting the number of training pixels in each class at each leaf node, such as:
  - Node 4: 80% forests, 20% agricultural land
  - Node 5: 60% agricultural land, 40% forests
- Assigning the leaf node to the class with the majority vote, such as:
  - Node 4: Forests
  - Node 5: Agricultural land

Step 5: Classification
In this step, we classify new pixels into different LULC classes by traversing the decision tree from the root node to the leaf node. This involves:

- Starting at the root node and traversing the tree based on the values of the predictor variables and reaching a leaf node and assigning the pixel to the corresponding LULC class, such as:
  - Node 4: Forests

Step 6: Accuracy Assessment
In this step, we evaluate the accuracy of the classified map by comparing it with the ground truth data.

3.2 Unsupervised ML Algorithms for LULC Classification

Unsupervised ML algorithms are used when the target class labels are unknown or unavailable. These algorithms identify patterns and structures in the data without prior knowledge of the class labels. Some popular unsupervised ML algorithms for LULC classification using GEE include:

- **K-Means Clustering**: A widely used algorithm that partitions the data into K clusters based on their similarities [14].
- **Hierarchical Clustering**: A algorithm that builds a hierarchy of clusters by merging or splitting existing clusters [14].

3.2.1 K-Means Clustering

K-Means clustering is an unsupervised algorithm that groups similar pixels into clusters based on their spectral characteristics. In the context of LULC classification, K-Means clustering can be used to identify different land cover classes such as forests, agricultural land, waterbodies, and urban areas. For example, with a Landsat 8 image of a region with a mix of forests, agricultural land, and urban areas. K-Means clustering can be used to group the pixels into 5 clusters based on their spectral characteristics. The algorithm will identify the following clusters:

Cluster 1: Water Bodies—Identified by low reflectance in near-infrared (NIR) and visible bands.
Cluster 2: Forests—High reflectance in NIR, low reflectance in red bands, and high NDVI values.
Cluster 3: Agricultural land—High reflectance in NIR, low reflectance in red bands and moderate NDVI values.
Cluster 4: Urban Areas—Reflectance patterns show high variability across multiple bands.
Cluster 5: Bare Soil—Moderate reflectance in visible and near infrared bands.

The resulting clusters can be used to create a LULC map, where each cluster is assigned a specific land cover class.

3.2.2 Hierarchical Clustering

Hierarchical clustering is another unsupervised algorithm that groups pixels into clusters based on their similarity. In GEE, you can use the ee.Clusterer.hierarchical() function to perform hierarchical clustering on a dataset.

4 Processes Involved in the LULC Classification Using Google Earth Engine

The Fig. 1 explains the processes involved in the LULC classification using GEE. It illustrates the sequence of steps, starting from data acquisition and preprocessing of Landsat 8 imagery to feature extraction and model training using the classifier.

5 Case Study

In this chapter, we present a case study on LULC change detection in the Upper Bhavani Basin using a supervised Random Forest classifier in GEE. This study demonstrates the application of supervised machine learning algorithms in GEE for LULC classification, highlighting their potential for land cover change detection and monitoring.

5.1 Description of the Study Area—Upper Bhavani Basin

The Upper Bhavani River Basin, a sub-basin of the Cauvery River Basin, is the focus of this study. Located in the western part of Tamil Nadu, India, it is bounded by Karnataka state to the north, the Coimbatore plateau to the south, the Bhavanisagar dam to the east, and Kerala state to the west. The basin covers an area of 7144 km^2 and has a length of

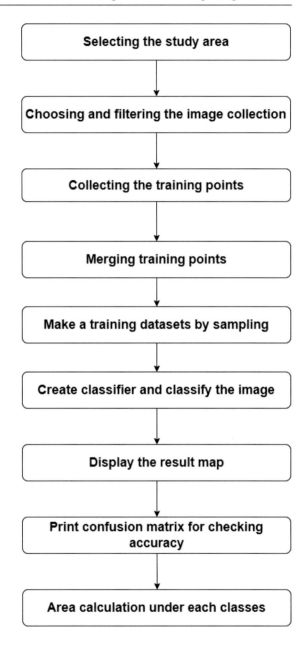

Fig. 1 Steps involved in LULC classification using Google Earth Engine

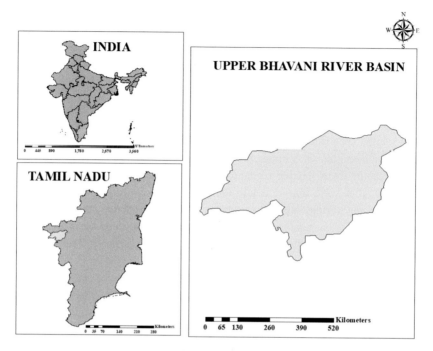

Fig. 2 Study area—upper Bhavani River Basin

216 km. The Bhavani Basin is further divided into two parts: the upper Bhavani and the lower Bhavani, with the upper part receiving heavier rainfall from both monsoons.

The climate in the Upper Bhavani River Basin is subtropical, with moderate temperatures. The summer season is hot and dry, with a maximum temperature of 40 °C, while the winter season is mild, with a minimum temperature of 22 °C. The topography of the Upper Bhavani River Basin is characterized by undulating terrain in the lower plateau and uplands, and rugged terrain in the western parts. The slope decreases towards the lower Bhavani (Fig. 2).

5.2 Methodology Adopted

Software and Data Used

In this study, the GEE platform is used, which is a cloud-based platform that allows users to access and analyse large amounts of remote sensing data without the need of downloading huge datasets. This enables efficient and scalable processing of data, reducing the computational resources required. The cloud-based architecture of GEE also provides a collaborative environment, which allows multiple users to work together on a project and share results in real-time. Furthermore, GEE's automated data processing and analysis capabilities enable rapid processing of large datasets, making it an ideal platform for large-scale remote sensing studies. The user interface of the GEE platform is given in the Fig. 3.

Data Used

In this study, Landsat 8 imageries acquired between 2017 and 2021 were imported using the Earth Engine Data Catalog, which provides access to a vast repository of satellite and aerial imagery. The Landsat 8 imageries were selected for their high spatial and temporal resolution, making them suitable for land cover classification and change detection analysis. The imageries were imported into the GEE platform, where they were processed and analysed using various algorithms and techniques. The five-year period of imagery acquisition allows for the analysis of land cover changes and trends over time.

Landsat 8 imagery is collected for the Upper Bhavani Basin in Tamil Nadu, India, for the period of 2017–2021, ensuring that the images are cloud-free and have minimal atmospheric interference. The imagery is imported into GEE and pre-processing steps are performed, including atmospheric correction using the Landsat 8 Surface Reflectance product, application of a cloud mask, and clipping to the Upper Bhavani Basin boundary.

A classification model is developed for the Upper Bhavani Basin, including classes such as forest/plantations, agriculture, urban, water bodies, barren land, and grasslands. The pre-processed imagery is split into training (80%) and testing (20%) datasets, and the RF classifier in GEE is used to train the model using the training dataset. Hyperparameter tuning is performed to optimize the RF model's performance.

Fig. 3 User interface of Google Earth Engine

The trained RF model is used to classify the entire Upper Bhavani Basin imagery for each year from 2017 to 2021, generating a LULC map for each year with each class represented by a unique colour or symbol. The testing dataset is used to evaluate the accuracy of the LULC classification, with metrics such as overall accuracy, producer's accuracy, user's accuracy, and kappa coefficient being calculated. A confusion matrix analysis is performed to identify errors and areas of improvement. A change detection analysis is performed to identify areas of LULC change between 2017 and 2021based on the area under each class.

6   Results and Discussion

The land cover classification and change detection analysis using Landsat 8 imageries acquired between 2017 and 2021 revealed significant changes in the study area (Figs. 4, 5, 6, 7 and 8). The results are presented in the Table 1.

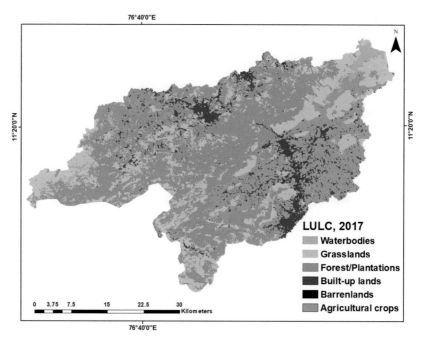

Fig. 4 Classified LULC map of the year 2017

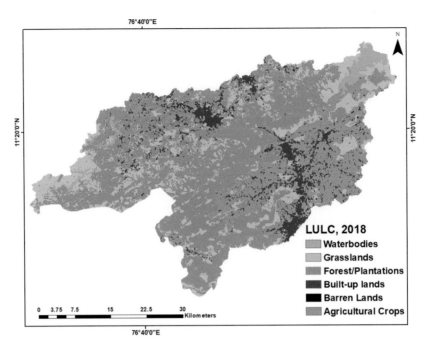

Fig. 5 Classified LULC map of the year 2018

6.1 Trends and Observations

- **Water bodies**: The area of water bodies has increased significantly from 2017 to 2020, with a peak in 2020, and then decreased slightly in 2021. This is due to the changes in rainfall in the study area (Table 1).
- **Forest/Plantation**: The area of forest/plantation has shown a fluctuating trend, with a decrease in 2019 and an increase in 2021. This could be due to factors such as deforestation, afforestation, or changes in forest management practices. Different nature conservation schemes and projects like MGNREGS (Mahatma Gandhi National Rural Employment Guarantee Act) has great role in this (Table 1).
- **Agricultural crops**: The area of agricultural crops has been decreasing steadily from 2017 to 2020, with a slight increase in 2021. This could be due to changes in agricultural practices, crop rotation, or land use conversion/urbanization (Table 1).
- **Built-up lands**: The area of built-up lands has been increasing steadily from 2017 to 2021, indicating urbanization and infrastructure development (Table 1).
- **Barren lands**: The area of barren lands has been very small and has shown a slight increase from 2019 to 2021. This could be due to changes in land use or land degradation (Table 1).

6 Results and Discussion

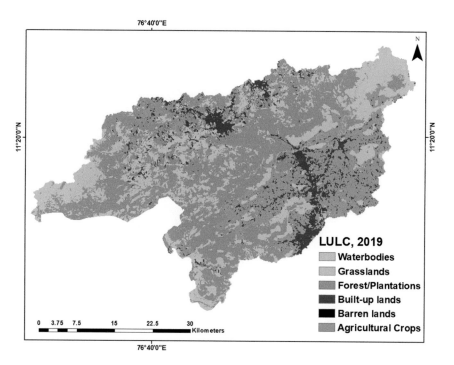

Fig. 6 Classified LULC map of the year 2019

- **Grasslands**: The area of grasslands has shown a fluctuating trend, with a decrease in 2018 and an increase in 2019, followed by a decrease in 2021. This could be due to factors such as overgrazing, land use conversion, or changes in vegetation cover (Table 1).

The results of the land cover classification and change detection analysis reveal significant changes in the study area between 2017 and 2021. The increase in water bodies and built-up lands suggests urbanization and infrastructure development in the area. The steady increase in forest and plantation areas may be attributed to afforestation and reforestation efforts under the different schemes like MGNREGS scheme. The decrease in agricultural crops and grasslands may be due to urbanization in the area.

The influence of different scheme activities and conservation practices has had a significant impact on the land cover changes in the study area. The implementation of these activities and practices has contributed to the improvement of environmental sustainability, livelihoods of rural communities, and ecosystem health.

Further analysis and investigation are required to understand the underlying causes of these changes and to develop strategies for sustainable land use planning and management. The integration of MGNREGS scheme activities and conservation practices into land use

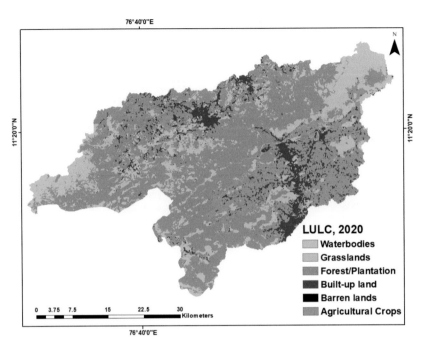

Fig. 7 Classified LULC map of the year 2020

planning and management can help to achieve sustainable development goals and improve the overall well-being of the environment and human populations in the study area.

6.2 Accuracy Assessment

Accuracy was assessed based on the over all accuracy, kappa coefficient values, producer's accuracy and user's accuracy (Tables 2, 3 and 4). Overall Accuracy represents the proportion of correctly classified pixels out of the total number of pixels. An overall accuracy of 90% or higher is generally considered acceptable [15].

Kappa Coefficient measures the agreement between the classified map and the reference data, accounting for chance agreement. A kappa coefficient of 0.8 or higher is generally considered acceptable [15].

The accuracy and kappa coefficient values obtained in this study indicate a high level of accuracy and reliability in the LULC map analysis. The producer's accuracy and user's accuracy values are also high, indicating that the classification of different land use/land cover classes is reliable.

6 Results and Discussion

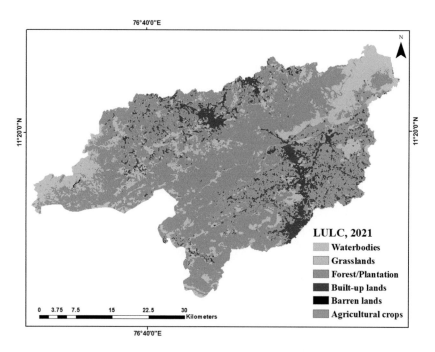

Fig. 8 Classified LULC map of the year 2021

Table 1 Changes in LULC over the periods

	2017	2018	2019	2020	2021
Water bodies	33.798	49.680	59.444	73.384	68.898
Forest/plantation	824.514	903.536	795.096	893.277	946.863
Agricultural crops	394.616	379.173	366.217	341.143	349.763
Built-up lands	140.109	141.766	153.311	161.774	170.604
Barren lands	1.580	0.056	0.090	0.415	0.697
Grasslands	439.997	360.401	460.455	364.620	297.788

Table 2 Overall accuracy and kappa coefficient

Year	Overall Accuracy (OA) (%)	Kappa coefficient (κ)
2017	91.23	0.92
2018	92.15	0.93
2019	93.05	0.94
2020	93.95	0.95
2021	94.85	0.96

Table 3 Users accuracy

Class	2017 (%)	2018 (%)	2019 (%)	2020 (%)	2021 (%)
Water bodies	90.48	92.31	93.85	95.38	96.15
Forest/plantation	95.74	96.49	97.22	97.95	98.67
Agricultural crops	90.91	92.86	94.44	95.83	96.92
Built-up lands	96.30	97.06	97.78	98.33	98.89
Barren lands	90.00	91.67	92.86	93.75	94.44
Grasslands	92.31	94.12	95.45	96.30	97.22

Table 4 Producers accuracy

Class	2017 (%)	2018 (%)	2019 (%)	2020 (%)	2021 (%)
Water bodies	96.50	97.50	98.50	99.00	99.50
Forest/plantation	94.50	95.50	96.50	97.00	97.50
Agricultural crops	92.50	93.50	94.50	95.00	95.50
Built-up lands	95.00	95.83	96.67	97.22	97.78
Barren lands	90.00	91.50	92.50	93.50	94.00
Grasslands	93.33	94.44	95.56	96.11	96.67

7 Conclusion

This chapter offers a thorough examination of LULC change detection methodologies, highlighting the integration of Remote Sensing, GIS, and Machine Learning techniques for modern environmental monitoring. Over time, LULC classification has transitioned from traditional manual methods to advanced machine learning algorithms, improving the precision and efficiency of landscape change analysis. Both supervised and unsupervised machine learning models have established significant potential in LULC mapping, each suited to different study requirements. The case study of the Upper Bhavani Basin illustrates the practical application of these methodologies, utilizing Landsat 8 imagery and the Random Forest classifier within the Google Earth Engine (GEE) platform. High overall accuracy and satisfactory Kappa coefficient values confirm the acceptability of the Random Forest model for LULC classification in this context. The observed changes in the Upper Bhavani Basin from 2017 to 2021 emphasize the need for continuous monitoring to sustainably manage natural resources. As environmental challenges grow more complex, the integration of advanced machine learning techniques and real-time data processing will be essential for improving the accuracy and timeliness of LULC assessments.

References

1. Chughtai, A. H., Abbasi, H., & Karas, I. R. (2021). A review on change detection method and accuracy assessment for land use land cover. *Remote Sensing Applications: Society and Environment*, 22, 100482.
2. Attri, P., Chaudhry, S., & Sharma, S. (2015). Remote sensing & GIS based approaches for LULC change detection–a review. *International Journal of Current Engineering and Technology*, 5(5), 3126–3137.
3. Kafi, K. M., Shafri, H. Z. M., & Shariff, A. B. M. (2014, June). An analysis of LULC change detection using remotely sensed data; A Case study of Bauchi City. In *IOP conference series: Earth and environmental science* (Vol. 20, No. 1, p. 012056). IOP Publishing.
4. Hamud, A. M., Shafri, H. Z. M., & Shaharum, N. S. N. (2021, May). Monitoring urban expansion and land use/land cover changes in banadir, somalia using google earth engine (GEE). In *IOP Conference Series: Earth and Environmental Science* (Vol. 767, No. 1, p. 012041). IOP Publishing.
5. Pande, C. B. (2022). Land use/land cover and change detection mapping in Rahuri watershed area (MS), India using the google earth engine and machine learning approach. *Geocarto International*, 37(26), 13860–13880.
6. Sellami, E. M., & Rhinane, H. (2023). Google Earth Engine and Machine Learning for Mapping Flash Flood Exposure-Case Study: Tetouan, Morocco
7. Wang, J., Bretz, M., Dewan, M. A. A., & Delavar, M. A. (2022). Machine learning in modelling land-use and land cover-change (LULCC): Current status, challenges and prospects. *Science of The Total Environment*, 822, 153559.
8. Liou, Y. A., & Rahman, A. (2020). Land-Use Land-Cover Classification by Machine Learning Classifiers for Satellite Observations—A Review. *Remote Sens*, 12, 1135.
9. Talukdar, S., Singha, P., Mahato, S., Pal, S., Liou, Y. A., & Rahman, A. (2020). Land-use land-cover classification by machine learning classifiers for satellite observations—A review. *Remote sensing*, 12(7), 1135.
10. Al-doski, J., Mansor, S. B., & Shafri, H. Z. (2013, April). Support vector machine classification to detect land cover changes in Halabja City, Iraq. In *2013 IEEE Business Engineering and Industrial Applications Colloquium (BEIAC)* (pp. 353–358). IEEE.
11. Kumari, A., & Karthikeyan, S. (2022, December). Comparative Performance of Maximum Likelihood and Minimum Distance Classifiers on Land Use and Land Cover Analysis of Varanasi District (India). In *International Conference on Advanced Network Technologies and Intelligent Computing* (pp. 476–484). Cham: Springer Nature Switzerland.
12. Kathar, S. P., Nagne, A. D., Awate, P. L., & Bhosle, S. (2023, December). Comparative Study of Supervised Classification for LULC Using Geospatial Technology. In *International Conference on Soft Computing and its Engineering Applications* (pp. 79–93). Cham: Springer Nature Switzerland.
13. Bayas, S., Sawant, S., Dhondge, I., Kankal, P., & Joshi, A. (2022). Land use land cover classification using different ml algorithms on sentinel-2 imagery. In *Advanced machine intelligence and signal processing* (pp. 761–777). Singapore: Springer Nature Singapore.
14. Vitale, A., Salvo, C., & Lamonaca, F. (2024, June). A Novel Geospatial Methodology for Measuring and Mapping Spatiotemporal Built-Up Dynamics Based on Google Earth Engine and Unsupervised K-Means Clustering of Multispectral Satellite Imagery. In *2024 IEEE International Workshop on Metrology for Living Environment (MetroLivEnv)* (pp. 57–62). IEEE.

15. Nasiri, V., Deljouei, A., Moradi, F., Sadeghi, S. M. M., & Borz, S. A. (2022). Land use and land cover mapping using Sentinel-2, Landsat-8 Satellite Images, and Google Earth Engine: A comparison of two composition methods. *Remote Sensing, 14*(9), 1977.

Watershed Prioritization for Rainwater Harvesting Using Multi-criteria Analysis, GIS, and Remote Sensing

Abstract

Rainwater harvesting is a promising tool for augmenting surface water and groundwater supply. This study has made an attempt to prioritize the Noyyal basin, Tamil Nadu, using GIS based weighted overlay analysis. Rainwater harvesting potential zones were delineated using the maps corresponding to slopes, drainage density, lineament density, soil, landuse, geology, rainfall and geomorphology. Each thematic layer was reclassified and suitable weightages were assigned to these layers and also to the different classes of each layer based on their relative influence on rainwater harvesting potential. The potential zone score was computed by using weighted overlay analysis. Rainwater harvesting potential zones in the study area are classified based on rainwater harvesting potential score. A high runoff potential area was designated for approximately 16 percent of the watershed.

Keywords

DEM • GIS • Rainwater harvesting • Weighted overlay

1 Introduction

Water is a fundamental natural resource essential for sustaining life and driving economic development. Its significance is not only limited to fulfilling basic human needs but also extends to supporting agriculture, industry, and ecosystems. In recent times, however, water scarcity has become a critical issue, exacerbated by factors such as rapid population growth, urbanization, industrial expansion, and inefficient management and conservation of water resources. Consequently, addressing water scarcity has emerged as a crucial challenge for many regions around the world [1, 2].

The demand for water is steadily rising due to an increasing global population and heightened levels of human activities. These pressures are further compounded by the continuous depletion of freshwater resources, both surface and groundwater, alongside the unpredictable impacts of climate change. In many parts of the world, including Tamil Nadu in southern India, water resources are becoming increasingly strained. Projections suggest that if the current trends continue, the per capita water availability in Tamil Nadu could fall to a mere 416 m^3 by 2050, representing a staggering deficit of 75.53% compared to current levels [3]. Such a shortfall could have severe repercussions, potentially reducing the availability of water to the extent that only one meal a day may be feasible for the state's population.

Rainfall serves as the primary natural source of freshwater replenishment. It is critical to capture and utilize this resource effectively to mitigate the effects of water scarcity. Unfortunately, a substantial portion of rainfall is often lost as surface runoff, quickly flowing out of catchments without being utilized, thereby reducing its potential benefits.

While rainwater harvesting plays a vital role in augmenting water resources, the identification and development of groundwater potential zones are equally crucial. Groundwater serves as a significant source of freshwater, especially during periods of low rainfall and drought. However, unregulated extraction and poor recharge practices have led to alarming declines in groundwater levels across many regions. Identifying areas with high groundwater potential can guide the strategic placement of rainwater harvesting structures, ensuring that collected rainwater is directed towards zones where it can effectively recharge aquifers [4, 5].

Utilizing tools like Remote Sensing (RS) and Geographic Information Systems (GIS) enables the integration of multiple data layers-such as topography, land use, soil type, and rainfall patterns—to identify and map groundwater potential zones accurately [5]. These maps are invaluable for planning effective groundwater recharge initiatives, optimizing the placement of recharge structures, and managing water resources sustainably. By focusing on these potential zones, water resource managers can enhance groundwater recharge, mitigate over-extraction, and ensure a more balanced use of surface and subsurface water resources [3]. Constructing RWH structures in areas identified as high groundwater potential zones can significantly enhance the efficiency of recharge processes, which also ensures that the harvested rainwater contributes effectively to replenishing depleted aquifers. This approach not only helps in meeting current water demands but also secures water availability for future generations, supporting long-term sustainability and resilience against climatic uncertainties.

2 Materials and Methods

2.1 Study Area—Noyyal River Basin

The Noyyal River originates in the Velliangiri Hills, often referred to as the southern Kailayam, situated in the Western Ghats. The river traverses a distance of approximately 180 km before it merges with the Cauvery River at Kodumudi. The Noyyal River Basin covers parts of Coimbatore, Erode, and Tiruchirappalli districts, covering a catchment area of about 3,510 km^2. The geographical boundaries of the basin lie between 10° 54′ 00″ to 11° 19′ 03″ North latitude and 76° 39′ 30″ to 77° 05′ 25″ East longitude [6].

The basin experiences significant spatial variability in annual rainfall, with the upper catchment areas receiving more than 3,000 mm of rainfall, whereas the lower basin regions receive as little as 600 mm annually. The pre-monsoon season, particularly summer rainfall during April and May, contributes around 100–300 mm of rainfall. Being a seasonal river, the Noyyal experiences substantial flow primarily during the North-East and South-West monsoon seasons. Outside these periods, the river typically has a scanty flow, with occasional flash floods occurring during heavy rains in the catchment areas [6].

The Noyyal River plays a crucial role in supporting agricultural activities within its basin. It supplies water to numerous irrigation tanks, especially around the Coimbatore area and downstream regions. Approximately 6,000 acres of cultivable land in Coimbatore district depend on the river for irrigation. However, the river basin has undergone significant environmental degradation due to rapid industrialization and urbanization. Thousands of small-scale industries, particularly in the textile and dyeing sectors, have been established around the Noyyal River Basin. Unfortunately, many of these industries discharge treated and untreated wastewater, along with sludge, directly into the river. Furthermore, sewage from the cities of Coimbatore and Tiruppur is often released into the river without adequate treatment, contributing to the severe pollution levels observed in the basin [6] (Fig. 1).

Despite its relatively small size compared to other Indian rivers, the Noyyal River exemplifies the ongoing struggle between resource utilization and environmental degradation. Parts of the basin are now characterized as "Industrial Wastelands," suffering from significant ecological damage due to the discharge of industrial effluents. Consequently, the groundwater in proximity to the river and the river water itself have become unfit for drinking, industrial, and irrigation purposes in many areas.

Given the environmental pressures faced by the Noyyal River Basin, identifying groundwater potential zones is essential for sustainable water resource management. Mapping these zones can help determine where groundwater can be effectively recharged and utilized, ensuring that water supplies remain reliable for both agricultural and industrial uses.

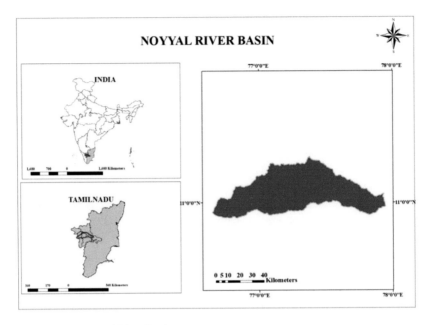

Fig. 1 Study area—Noyyal River Basin

2.2 Data Collection

a. **Base Map**

The base map of the Noyyal River Basin was delineated using data from the Shuttle Radar Topography Mission (SRTM) Digital Elevation Model (DEM), downloaded from the USGS explorer, which was processed using ArcGIS software. This DEM data enabled accurate delineation of the basin boundaries, capturing the topography and flow patterns essential for hydrological analysis.

b. **Slope Map**

The slope map was prepared from the SRTM DEM using spatial analyst tool in ArcGIS software. First, the DEM was processed using the 'Slope' function available in ArcGIS, which calculates the slope for each cell in the raster and expresses it in degrees. Using the reclassification tool in the Spatial Analyst extension, the slope values were categorized into different slope classes, each assigned a specific rank based on its suitability for groundwater recharge. The slope in the study area ranges from 0 to 55.88 degrees (Fig. 2) and are categorized into five classes: (a) very gentle (0°-5°), (b) gentle (5°-10°),

(c) moderate (10°-15°) (d) moderately steep (15°-20°) and (e) steep (>20°). The slopes less than 5% is more suitable for constructing rainwater harvesting structures [7].

c. **Drainage Network Map**

The drainage network for the study area was generated by extracting stream data from the SRTM DEM (30 m), following standard hydrological modelling techniques within GIS. The accuracy of the derived stream network was further validated through a ground-truth survey conducted across the basin to ensure that the modelled streams aligned with the actual streams and water bodies present on the ground.

d. **Drainage Density Map**

Drainage density, defined as the total length of streams per unit area, is a key factor influencing surface runoff and groundwater recharge. The drainage network was extracted and subsequently converted to vector format. The drainage density map was finally created using the 'Line Density' tool, which calculates the length of all streams within a specified area. The map was classified into various density classes as shown in Table 1.

e. **LULC and Soil Map**

In terms of spatial data layers, the Land Use and Land Cover (LULC) map and the soil map were acquired from the Department of Remote Sensing and Geographical Information System (GIS) at Tamil Nadu Agricultural University (TNAU), Coimbatore. These maps provide essential information about the spatial distribution of land cover types and soil properties, both of which significantly influence groundwater recharge potential and surface runoff characteristics in the basin. The soil map and LULC maps, originally in vector format (shapefile), were obtained and converted to raster format using the polygon-to-raster conversion tool in ArcGIS. These soil and LULC classes were reclassified to simplify the analysis and ranks were assigned to each soil type based on its permeability and potential for groundwater recharge (as per the Table 1).

f. **Rainfall**

Rainfall data, a key component for hydrological modelling and water resource assessment, was gathered from the Tamil Nadu Surface and Groundwater Data Centre, Public Works Department (PWD), Taramani, Chennai. Eleven rain gauge stations, located in proximity to the Noyyal River Basin, were identified for detailed analysis (Table 2). The data collected from these stations covers the period from 2013 to 2019 and is maintained in Excel format. This temporal rainfall data was projected in the GIS environment, allowing for spatial interpolation to understand rainfall distribution across the basin, which is

Table 1 Weights assigned for different ground water control factors

Thematic layer	Class	Area, km²	Class score	Weight	Total score
Geology	Fluvial/Coastal and Glacial Sediments	265.021	7	6	42
	Acid Intrusive/Granite/Granodiorite	253.221	6		36
	Alkali Complex Gp.	11.223	5		30
	Khondalite Gneissic Complex	19.884	4		24
	Peninsular Gneissic Complex-1	960.884	3		18
	Charnockite Gneissic Complex	85.344	2		12
	Migmatite Gneissic Complex	1944.778	1		6
Geomorphology	Waterbodies	52.947	6	14	84
	Flood Plain	0.965	6		84
	Pediment Pedi plain Complex	3184.586	5		70
	Highly Dissected Hills and Valleys	188.393	4		56
	Low Dissected Hills and Valleys	7.919	3		42
	Bajada	83.838	2		28
	Quarry and Mine Dump	21.669	1		14
Land use	Waterbodies	34.7106	6	12	72
	Evergreen forest	65.275	5		60
	Mixed forest	21.8767	4		48
	Deciduous forest	83.7208	4		48
	Shrub land	40.0033	3		36
	Plantations	118.951	3		36
	Grassland	239.098	3		36
	Cropland	1244.41	3		36
	Wasteland	7.14472	2		24
	Fallow land	1234.23	2		24
	Barren land	3.50293	2		24
	Built-up land	411.679	1		12

(continued)

Table 1 (continued)

Thematic layer	Class	Area, km²	Class score	Weight	Total score
Soil	Sand	6.8562	8	8	64
	Loamy sand	110.074	7		56
	Sandy loam	568.66	6		48
	Sandy clay loam	1037.78	5		40
	Sandy clay	290.141	4		32
	Clay loam	784.23	3		24
	Clay	523.633	2		16
	Built-up land	218.818	1		8
Lineament density	High	230.041	3	14	42
	Medium	189.077	2		28
	Low	3121.14	1		14
Drainage density	Very high	22.3736	4	12	48
	High	557.071	3		36
	Medium	1694.96	2		24
	Low	22.3736	1		12
Rainfall	High	109	4	14	56
	Moderate	650	3		42
	Low	2601	2		28
	Very Low	180.32	1		14
Slope	Very Gentle	969.521	5	18	90
	Gentle	1711.91	4		72
	Moderate	561.085	3		54
	Moderately steep	124.741	2		36
	Steep	173.063	1		18

crucial for identifying areas with higher recharge potential and managing water resources effectively.

To create the rainfall map, the Inverse Distance Weighted (IDW) interpolation technique was used in ArcGIS. This method was applied to generate a spatial distribution of rainfall across the study area using averaged rainfall data from 11 gauging stations over a period of seven years (2013–2019). The process began by inputting the average rainfall values from each of the 11 gauging stations into the GIS environment. The IDW interpolation technique was then employed to estimate rainfall values for the entire study area. IDW assigns more influence to the points closer to the location of interest, making it an effective tool for depicting rainfall variability within the basin. The obtained rainfall

Table 2 Rain gauge stations near to Noyyal basin

No.	Station	Longitude	Latitude
1	Annur	77.10639	11.23278
2	Chithraichavadi Anicut	76.82639	10.96722
3	Coimbatore Airport	77.02361	11.02611
4	Agriculture College	76.935	11.01222
5	CMC Medical College	77.02611	11.02972
6	Periyanayakkanpalayam	76.94639	11.15222
7	Pothanur railway	76.98917	10.96472
8	Sultanpet	77.19639	10.87528
9	Sulur	77.12278	11.025
10	Thiruppur	77.34389	11.10472
11	Thondamuthur	76.84306	10.98833

map was categorized into different classes, each assigned a specific rank based on its suitability for groundwater recharge (Table 1).

g. **Lineament Density**

The lineament map (which highlights geological fractures and faults that may influence groundwater movement) was sourced from the Bhukosh platform (https://bhukosh.gsi.gov.in/Bhukosh/Public), managed by the Geological Survey of India (GSI). The map was projected using a consistent coordinate system to ensure accurate overlay and analysis in GIS. Lineaments, representing fractures and faults in the earth's crust, are crucial indicators of groundwater movement, especially in hard rock terrains. The lineament density map was created by first mapping the lineaments and then calculating the lineament density using the 'Line Density' tool in ArcGIS. Lineament density is expressed as the total length of lineaments per unit area. This map was further classified into 3 categories: Low, Medium, and High. Areas with higher lineament density, indicating higher secondary porosity and permeability, were assigned higher ranks due to their potential for enhanced groundwater recharge.

Additional critical datasets, such as geology, and geomorphology maps, were sourced from the Bhukosh platform (https://bhukosh.gsi.gov.in/Bhukosh/Public), managed by the Geological Survey of India (GSI). These maps were projected using a consistent coordinate system to ensure accurate overlay and analysis in GIS. The information from these maps is pivotal for identifying groundwater potential zones, as lineaments can often serve as conduits for groundwater flow, and geomorphological features can dictate the terrain's capacity for water storage and infiltration. The geomorphological and geological units were reclassified based on their groundwater potential, with each unit being assigned a rank according to its ability to support groundwater recharge (Table 1).

By integrating all these spatial and temporal data layers—topography, drainage network, LULC, soil, geology, lineaments, geomorphology, and rainfall—into a Geographic Information System (GIS) framework, a complete analysis of the Noyyal River Basin can be undertaken. The consistent use of projection systems across all datasets ensures spatial accuracy and alignment, which is critical when determining groundwater potential zones, planning rainwater harvesting structures, and assessing overall water resource management in the basin.

By projecting all thematic layers to the WGS 1984 UTM Zone 43N coordinate system, a consistent spatial reference framework was established for the entire basin.

2.3 Data Integration and Analysis in GIS Environment

The integration of thematic layers is performed using the weighted overlay analysis within the ArcGIS environment. During this process, each thematic layer—such as geology, slope, land use/land cover, geomorphology, drainage density, soil type, and lineament density—is assigned a weight in the weighted overlay table, to reflect its relative influence on groundwater potential. The classes within each layer are then ranked accordingly. Subsequently, the weighted overlay analysis is conducted in ArcMap, combining these layers based on their assigned weights. The resulting integrated map delineates the study area into different classes, representing varying degrees of groundwater potential zones. This classification provides a visual representation of areas with high, moderate, and low groundwater prospects, supporting effective decision-making for groundwater exploration and management (Table 3).

Table 3 Recharge potential categories and criteria

S. No.	Class name	Criteria	Potential zone score range	Area (km^2)
1	Very high potential	$> (\mu + \sigma)$	297.65–318.00	555.846
2	High potential	μ to $(\mu + \sigma)$	270.00–297.65	1168.65
3	Moderate potential	$(\mu - \sigma)$ to μ	242.31–270.00	1177.41
4	Low potential	$< (\mu - \sigma)$	118.00–242.31	582.147

3 Results and Discussion

The various thematic layers used in the study were briefly discussed in this section below:

Eight thematic layers were used to prepare groundwater potential zone map of the study area. The area under each class and corresponding weights and total scores were given in the Table 1.

3.1 Slope Map

The slope map of the study area, as illustrated in Fig. 2, categorizes the terrain into various slope classes, each of which plays a significant role in determining groundwater potential zones [8]. The distribution is as follows: very gentle (969.521 km^2), gentle (1711.91 km^2), moderate (561.085 km^2), moderately steep (124.741 km^2), and steep (173.063 km^2).

Very Gentle Slopes (0–5% slope): Covering the largest area (969.521 km^2), these slopes are most conducive to groundwater recharge. The low gradient allows water to percolate into the soil, minimizing surface runoff and promoting infiltration. Areas with very gentle slopes are thus identified as high groundwater potential zones.

Gentle Slopes (5–10% slope): Spanning 1711.91 km^2, gentle slopes also support groundwater recharge, though to a slightly lesser extent than very gentle slopes. They have moderate infiltration rates, making them suitable for groundwater recharge but with a higher chance of surface runoff compared to very gentle slopes. These areas are generally considered moderate to high potential zones (Fig. 2).

Moderate Slopes (10–15% slope): With an area of 561.085 km^2, moderate slopes have reduced infiltration capacity as compared to gentler slopes. The increased gradient facilitates more surface runoff, which can lead to reduced groundwater recharge. These

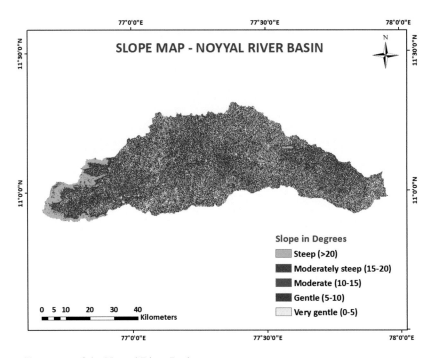

Fig. 2 Slope map of the Noyyal River Basin

3 Results and Discussion

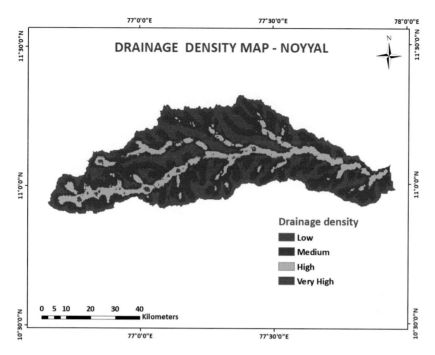

Fig. 3 Drainage density map of the Noyyal River Basin

regions are typically classified as moderate groundwater potential zones, where recharge is possible but less efficient.

Moderately Steep Slopes (15–20% slope): Covering 124.741 km², these slopes experience significant surface runoff, which greatly limits infiltration and groundwater recharge. The steeper gradient increases the velocity of runoff, decreasing the contact time for water to infiltrate into the ground. Consequently, these areas are categorized as low groundwater potential zones.

Steep Slopes (>20% slope): Representing the smallest area (173.063 km²), steep slopes are characterized by high surface runoff and minimal infiltration. The steep gradient facilitates rapid runoff, leaving little opportunity for groundwater recharge. These areas are therefore identified as very low groundwater potential zones and are generally unsuitable for groundwater recharge interventions.

3.2 Drainage Density Map

The drainage density map in Fig. 3 categorizes the study area into very high (22.4 km²), high (557.071 km²), medium (1694.96 km²), and low (22.3736 km²) density zones, each reflecting distinct groundwater potential characteristics. Areas with high drainage density

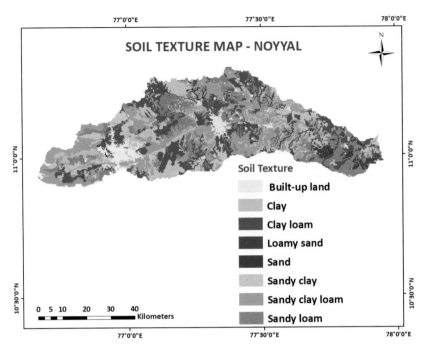

Fig. 4 Soil map of the Noyyal River Basin

are indicative of well-developed stream networks, which often correlate with enhanced groundwater recharge potential.

3.3 Soil Map

The soil textures can be categorized into three groups based on their infiltration rates and water-holding capacities. Sand (6.8562 km^2) and loamy sand (110.074 km^2) have high infiltration rates, providing the highest potential for groundwater recharge. Sandy loam (568.66 km^2) and sandy clay loam (1037.78 km^2) exhibit moderate infiltration rates and moderate recharge potential. In contrast, sandy clay (290.141 km^2), clay loam (784.23 km^2), and clay (523.633 km^2) have low infiltration rates and, consequently, lower potential for groundwater recharge [8] (Fig. 4).

3.4 LULC Map

The land use types can be categorized into three groups based on their potential for groundwater recharge. Waterbodies (34.7106 km^2) and evergreen forest (65.275 km^2)

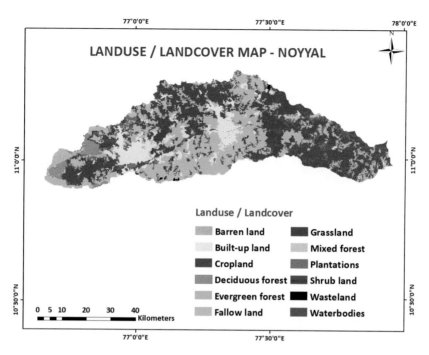

Fig. 5 LULC map of the Noyyal River Basin

have moderate recharge potential (class score 6 and 5, respectively) due to their ability to facilitate infiltration and maintain soil moisture. Mixed forest (21.8767 km^2) and deciduous forest (83.7208 km^2) also contribute to recharge but with slightly lower potential (Class score of 4). Shrub land (40.0033 km^2), plantations (118.951 km^2), grassland (239.098 km^2), and cropland (1244.41 km^2) have limited recharge potential (Class score of 3) due to reduced infiltration rates. Wasteland (7.14472 km^2), fallow land (1234.23 km^2), and barren land (3.50293 km^2) show very low recharge potential (class score of 2) because of poor soil structure and low infiltration. Built-up land (411.679 km^2), with its impermeable surfaces, has the least potential for groundwater recharge (Class score of 1) [8] (Fig. 5).

3.5 Rainfall

Based on annual average rainfall and its impact on groundwater potential, areas receiving less than 500 mm of rainfall (180.32 km^2) fall under the very low potential category. These areas are highly vulnerable to low groundwater recharge due to insufficient rainfall. Regions with 500–700 mm of rainfall (2601 km^2) have low groundwater potential, as limited rainfall restricts effective recharge, despite covering the largest area. Areas with

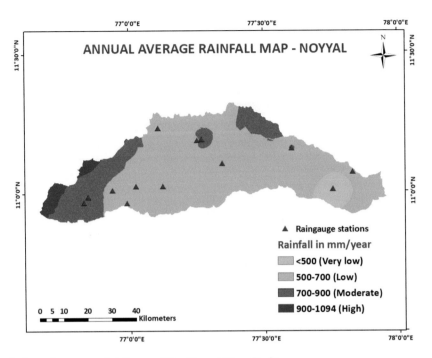

Fig. 6 Annual average rainfall map of the Noyyal River Basin

700–900 mm of rainfall (650 km^2) show moderate potential for groundwater recharge, benefiting from a balance of adequate rainfall and suitable conditions for infiltration. High-potential zones (109 km^2) receive 900–1094 mm of rainfall, promoting optimal groundwater recharge due to abundant rainfall, though these areas cover the smallest extent (Fig. 6) [9].

3.6 Geology Map

The Migmatite Gneissic Complex covers the largest area (1944.778 km^2) and has the lowest groundwater potential due to its low permeability and limited capacity for groundwater storage. The Charnockite Gneissic Complex (85.344 km^2) and the Peninsular Gneissic Complex-1 (960.884 km^2) follow, with slightly better groundwater potential but still constrained by their geological characteristics. The Khondalite Gneissic Complex (19.884 km^2) and the Alkali Complex Group (11.223 km^2) have moderate potential for groundwater recharge, although their limited area reduces their overall impact. The Acid Intrusive/Granite/Granodiorite formation (253.221 km^2) has relatively better groundwater potential due to the presence of fractured zones that aid recharge. The Fluvial/Coastal and Glacial Sediments (265.021 km^2) show the highest groundwater potential, as these unconsolidated

3 Results and Discussion

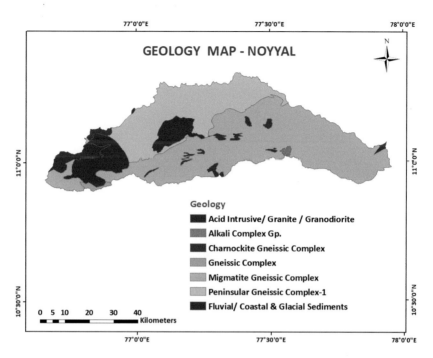

Fig. 7 Geology map of the Noyyal River Basin

deposits typically allow for higher permeability and better storage capacity. Prioritizing these geological formations can guide effective groundwater management and exploration strategies (Fig. 7) [10].

3.7 Geomorphology Map

The Quarry and mine dumps (21.669 km^2) have the lowest potential for groundwater recharge due to disturbed surfaces and compacted soil, which hinder effective infiltration. Bajadas (83.838 km^2) also have limited recharge potential, as these gently sloping depositional surfaces can only moderately facilitate water infiltration. Low dissected hills and valleys (7.919 km^2) offer slightly better potential compared to their highly dissected counterparts but still face challenges due to surface runoff and reduced infiltration. Highly dissected hills and valleys (188.393 km^2), with steep slopes and rapid runoff, provide low groundwater recharge potential. The pediment and pediplain complex (3184.586 km^2), covering the largest area, has moderate groundwater potential due to its gently sloping and weathered surfaces, which support some infiltration. Waterbodies (52.947 km^2) and flood plains (0.965 km^2) show the highest groundwater potential, as they naturally retain

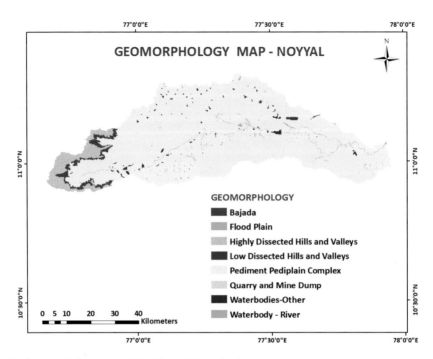

Fig. 8 Geomorphology map of the Noyyal River Basin

water and contribute significantly to groundwater recharge through their high permeability and water-holding capacity (Fig. 8) [10].

3.8 Lineament Density

Areas with high lineament density (230.041 km^2) have the greatest potential for groundwater recharge, as the well-developed network of fractures and faults facilitates enhanced infiltration and groundwater movement. These zones should be prioritized for groundwater exploration and sustainable management due to their capacity to effectively recharge aquifers. Medium lineament density areas (189.077 km^2) show moderate groundwater potential, with fewer fractures compared to high-density zones but still providing reasonable pathways for groundwater storage. Targeted recharge interventions and management practices can enhance their potential further. In contrast, low lineament density areas (3121.14 km^2) have the least groundwater recharge potential, covering the largest extent. The sparse network of fractures and faults in these areas limits water infiltration and movement, making them less favourable for recharge (Fig. 9) [11].

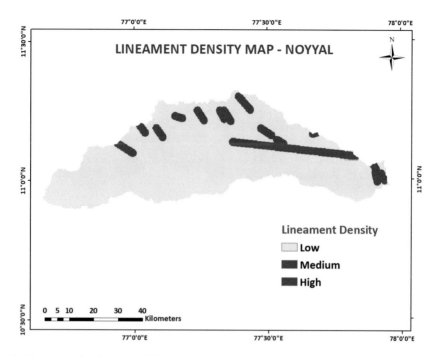

Fig. 9 Lineament density map of the Noyyal river Basin

3.9 Groundwater Potential Zone Map

In the current study, groundwater potential zones were classified based on specific criteria, as shown in Table 3. This analysis identifies four distinct categories of groundwater potential, reflecting the varying capacity for recharge across different areas. **Very high potential areas** cover 555.846 km^2 and represent regions with optimal conditions for groundwater recharge. **High potential areas**, encompassing 1168.65 km^2, indicate substantial potential for recharge but are not as pronounced as very high potential zones. **Moderate potential areas** span 1177.41 km^2 and show reasonable recharge potential under favourable conditions. Lastly, **low potential areas** cover 582.147 km^2 and suggest limited groundwater recharge capacity due to less favourable conditions. This classification helps prioritize areas for groundwater management and conservation efforts based on their recharge potential (Figs. 10 and 11).

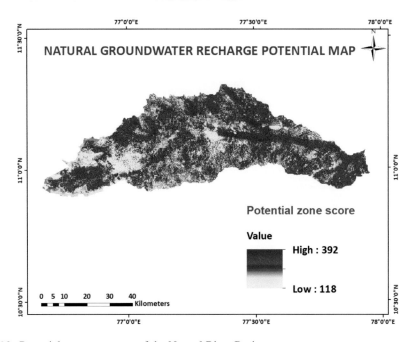

Fig. 10 Potential zone score map of the Noyyal River Basin

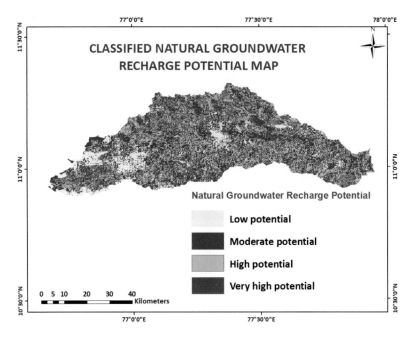

Fig. 11 Classified natural groundwater recharge potential map of the Noyyal River Basin

4 Conclusion

This study highlights the use of GIS-based weighted overlay analysis to prioritize rainwater harvesting potential zones in the Noyyal basin in Tamil Nadu. By examining various thematic layers including slope, drainage density, lineament density, soil, land use, geology, rainfall, and geomorphology, the study delineates areas suitable for rainwater harvesting. The rainwater harvesting potential zones were classified based on a harvesting potential score, revealing that approximately 16% of the watershed is designated as a high runoff potential area. This analysis highlights the importance of integrating multiple environmental factors in identifying effective rainwater harvesting sites.

References

1. Bagatin, R., Klemeš, J. J., Reverberi, A. P., & Huisingh, D. (2014). Conservation and improvements in water resource management: a global challenge. *Journal of Cleaner Production, 77,* 1–9.
2. Cook, H. F. (2017). *The protection and conservation of water resources.* John Wiley & Sons.
3. Natarajan, P. M., Kallolikar, S., Rengaraju, G., & Sambandam, S. T. (2013). Sustainable water resources development through integrated water management in Tamil Nadu, India. 86th Annual Water Environment Federation Technical Exhibition and Conference, WEFTEC 2013, 11, 7209–7217. https://doi.org/10.2175/193864713813726885.
4. Srivastava, R. C. (2001). Methodology for design of water harvesting system for high rainfall areas. *Agricultural Water Management, 47*(1), 37–53.
5. Singh, L. K., Jha, M. K., & Chowdary, V. M. (2017). Multi-criteria analysis and GIS modeling for identifying prospective water harvesting and artificial recharge sites for sustainable water supply. *Journal of cleaner production, 142,* 1436–1456.
6. Selvarani, A. G., Maheswaran, G., & Elangovan, K. (2017). Identification of artificial recharge sites for Noyyal River Basin using GIS and remote sensing. *Journal of the Indian Society of Remote Sensing, 45,* 67–77.
7. Moharir, K. N., Pande, C. B., Gautam, V. K., Singh, S. K., & Rane, N. L. (2023). Integration of hydrogeological data, GIS and AHP techniques applied to delineate groundwater potential zones in sandstone, limestone and shales rocks of the Damoh district, (MP) central India. *Environmental research, 228,* 115832.
8. Fatema, K., Joy, M. A. R., Amin, F. R., & Sarkar, S. K. (2023). Groundwater potential mapping in Jashore, Bangladesh. *Heliyon, 9*(3).
9. Baghel, S., Tripathi, M. P., Khalkho, D., Al-Ansari, N., Kumar, A., & Elbeltagi, A. (2023). Delineation of suitable sites for groundwater recharge based on groundwater potential with RS, GIS, and AHP approach for Mand catchment of Mahanadi Basin. *Scientific Reports, 13*(1), 9860.
10. Etuk, M. N., Igwe, O., & Egbueri, J. C. (2023). An integrated geoinformatics and hydrogeological approach to delineating groundwater potential zones in the complex geological terrain of Abuja, Nigeria. *Modeling Earth Systems and Environment, 9*(1), 285–311.
11. Arumugam, M., Kulandaisamy, P., Karthikeyan, S., Thangaraj, K., Senapathi, V., Chung, S. Y., ... & Manimuthu, S. (2023). An assessment of geospatial analysis combined with AHP techniques to identify groundwater potential zones in the Pudukkottai District, Tamil Nadu, India. *Water, 15*(6), 1101.

Study on Rainfall Variability and Distribution in Lower Bhavani River Basin, Tamil Nadu

Abstract

This study investigates the rainfall distribution patterns in the Lower Bhavani river basin, analysing annual, seasonal, and decadal variations using long-term daily rainfall data from 18 gauging stations (1981–2019), sourced from the Tamil Nadu Surface and Groundwater Data Centre, PWD, Taramani. The spatial distribution of rainfall across different scales was mapped using the Inverse Distance Weighted (IDW) interpolation method in the ArcGIS environment. The basin's average annual rainfall is approximately 722.8 mm, which is lower than the state-wide average for Tamil Nadu. The majority of rainfall occurs during the South-West and North-East monsoons, highlighting a pronounced seasonal variability within the basin. The study further identifies years with scanty, deficient, normal, and excess rainfall across different seasons. Decadal analysis shows that the average annual rainfall was 921 mm for the first decade (1981–1990), 936 mm for the second (1991–2000), 614.6 mm for the third (2001–2010), and 643 mm for the last decade (2011–2019). A shift in seasonal rainfall patterns was observed, with the South-West Monsoon dominating the early decade, while a shift towards the North-East Monsoon was noted in the later decades.

Keywords

IDW • Decadal rainfall distribution • Scanty rainfall • Annual average rainfall

1 Introduction

Rainfall variability is a critical aspect of the hydrological cycle, and its alteration has a direct impact on water resources. The hydrological cycle is a complex process that involves the continuous movement of water on, above, and below the surface of the Earth.

Rainfall is a vital component of this cycle, and any changes in its pattern can have far-reaching consequences for water resources. The changing pattern of rainfall, influenced by climate change, has been a concern for water resource managers and hydrologists [1, 2]. Climate change is altering the global climate, leading to changes in temperature, precipitation, and other weather patterns. These changes are, in turn, affecting the frequency, intensity, and distribution of rainfall events [3]. Researches have shown that changes in rainfall quantities and frequencies can alter stream flow patterns, groundwater reserves, and soil moisture, leading to widespread consequences on water resources, the environment, and agricultural productivity [4].

Some of the specific consequences of rainfall variability include:

- **Changes in Stream Flow Patterns**: Changes in rainfall patterns can alter the flow through rivers and streams, affecting the availability of water for irrigation, drinking water supply, and hydroelectric power generation [4].
- **Impact on Groundwater Reserves**: Rainfall variability can affect the recharge of groundwater aquifers, leading to changes in groundwater levels and quality. This can have significant implications for agriculture, industry, and drinking water supply [5].
- **Soil Moisture and Agricultural Productivity**: Changes in rainfall patterns can affect soil moisture levels, leading to changes in crop yields, food security, and agricultural productivity [6].
- **Environmental Consequences**: Rainfall variability can also have environmental consequences, such as changes in water quality, increased risk of flooding and landslides, and impacts on ecosystems and biodiversity [6].

Understanding rainfall variability is crucial for sustainable water resource management and agricultural development. Water resource managers and hydrologists need to understand the patterns and trends of rainfall variability to make informed decisions about water resource allocation, infrastructure development, and agricultural planning. By understanding the impacts of rainfall variability, we can develop strategies to mitigate its effects and ensure sustainable water resource management and agricultural development.

This study aims to investigate the rainfall variability in the Lower Bhavani Basin to provide insights for water resource management and agricultural planning.

2 Study Area—Lower Bhavani

The Bhavani River, the largest tributary of the Cauvery River, spans a length of 216 km and drains an area of approximately 7,144 km^2. The climate within the Bhavani basin, particularly in the upper reaches, is predominantly subtropical. This region has moderate temperatures, though the climate varies significantly across different seasons.

In the summer months, the region tends to experience hot and dry conditions, with temperatures rising as high as 40 °C. In contrast, during the winter months of November and December, temperatures drop, reaching a minimum of around 22 °C [7]. This fluctuation between extreme heat and milder conditions highlights the basin's diverse climatic patterns.

The primary source of water for the Bhavani River is the rainfall, which is influenced by weather systems such as depressions and cyclones, particularly during the two monsoon periods - the south-west monsoon (June to September) and the north-east monsoon (October to December). Additionally, localized thunderstorms during the hot season contribute to the river's water flow, although in smaller amounts.

The Bhavani basin is typically divided into two main sections: the upper Bhavani and the lower Bhavani. The upper Bhavani region is characterized by abundant rainfall, benefiting from both monsoons, making it relatively lush and fertile. However, as one moves towards the lower Bhavani, the amount of rainfall significantly diminishes. This results in a much drier landscape, especially during the winter and summer months when precipitation is minimal.

The Lower Bhavani basin, situated in Tamil Nadu, spans across parts of the Erode, Coimbatore, and Tirupur districts. Covering an area of 2,402 km^2, it lies between latitudes 11°15' and 11° 45' North and longitudes 77°0' and 77° 40' East, with ground elevations ranging from 154 to 1,669 m above mean sea level. The basin slopes toward the south-east, with the hilly regions covered by evergreen and deciduous forests, as well as plantations, while the plains are predominantly used for agriculture. The primary crops in the area include banana, paddy, and sugarcane, which are water-intensive (Fig. 1).

The climate is classified as semi-arid, with temperatures ranging from 17 °C to 38 °C. The soil composition varies throughout the basin, with coarser soils found in the hilly areas and finer, clay-rich soils in the low-lying regions near rivers and watercourses. The soil types include clay, clay loam, sandy clay, sandy loam, and silty clay, reflecting the diverse agricultural potential of the region.

3　　Rain Gauge Stations

Rainfall data from 18 gauging stations, covering the period from 1981 to 2019, were obtained from the Tamil Nadu Surface and Groundwater Data Centre, PWD, Taramani, to examine spatial and temporal rainfall variability. The rain gauge stations are more densely concentrated in the central and eastern parts of the basin, allowing for a more detailed analysis of rainfall patterns in these regions. The locations of the stations are illustrated in Fig. 2. The spatial arrangement and station density significantly impact the study of rainfall variability, as noted by Lee et al. [8]. Of the 18 stations, 13 are situated within the basin, while 5 are located outside its boundaries.

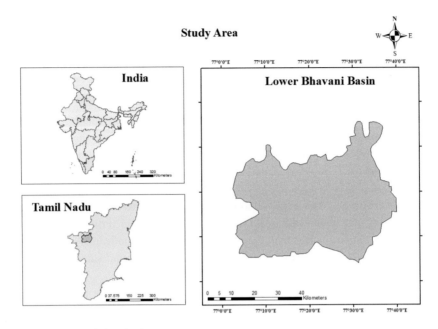

Fig. 1 Lower Bhavani river basin

Fig. 2 Locations of rain gauge stations

Table 1 Rainfall seasons in the study area

Sl. No.	Months	Season
1	January–February	Winter
2	March–May	Summer
3	June–September	South-West Monsoon (SWM)
4	October–December	North-East Monsoon (NEM)

4 Rainfall Distribution in the Lower Bhavani Basin

Rainfall in the Lower Bhavani basin follows a distinct seasonal pattern (Table 1), with the majority of the precipitation occurring during the North-East Monsoon, similar to the overall climate of Tamil Nadu. About 50% of the annual rainfall is received during this period, although the distribution tends to be irregular and inconsistent. Both spatial and temporal variations in rainfall are significant across the basin, leading to disparities in water availability in different areas.

5 Methodology for Estimating Average Annual, Seasonal, and Decadal Rainfall Using Daily Rainfall Data

To estimate the average annual rainfall, seasonal rainfall, and decadal average from daily rainfall data, a systematic approach is employed, utilizing Inverse Distance Weighted (IDW) interpolation and data cleaning processes. The following steps outline the methodology.

a. **Data Collection and Preparation**
   - *Data Source*: Daily rainfall data from 18 gauging stations in the Lower Bhavani basin were collected for the period from 1981 to 2019.
   - *Data Cleaning*: Eliminated the all daily rainfall values below 2.5 mm to exclude insignificant or trace rainfall amounts that do not contribute meaningfully to hydrological processes.
   - *Temporal Grouping*: Organized the daily data by year, season, and decade to facilitate analysis.

b. **Definition of Temporal Periods**
   - *Annual Rainfall*: Aggregated the daily rainfall data to calculate total and annual rainfall for each station.
   - *Seasonal Rainfall*: Defined four distinct seasons—Winter (Jan–Feb), Summer (Mar–May), Southwest Monsoon (Jun–Sep), and Northeast Monsoon (Oct–Dec). Summed up the daily rainfall for each season.

- **Decadal Rainfall**: Grouped the data into decadal intervals (e.g., 1981–1990, 1991–2000, etc.) and calculated the average rainfall for each decade.

c. **Spatial Analysis Using IDW Interpolation**
  - *Spatial Interpolation*: Applied IDW interpolation to estimate rainfall at unmeasured locations across the basin. This spatial analysis accounts for the variation in station density and location. Geographic Information System (GIS) software named ArcGIS 10.8 was used for spatial interpolation and visualization of the rainfall distribution across the basin.

d. **Calculating Averages**
  - *Annual Average Rainfall*: For each station, calculated the annual average rainfall by summing daily rainfall values over each year and dividing by the number of years.
  - *Seasonal Average Rainfall*: For each season, summed up the daily rainfall values during the defined seasonal periods and computed the average across all years.
  - *Decadal Average Rainfall*: Aggregated rainfall totals for each decade and computed the decadal average by dividing the total rainfall by the number of years in the decade.

e. **Spatial Averaging**
  - After interpolation, calculated the basin-wide spatial averages for annual, seasonal, and decadal rainfall by averaging the interpolated values across the entire basin grid.

6 Results and Discussion

The average annual rainfall represents the mean amount of precipitation recorded at a particular location over a span of thirty years. For this calculation, data from 1981 to 2019 were used, collected from 18 different rain gauge stations. The IDW interpolation method, as detailed by Yang et al. [9] was employed to compute the annual rainfall averages. Figure 3 illustrates the spatial distribution of the average annual rainfall across the lower Bhavani river basin.

The Lower Bhavani river basin receives an average annual rainfall of approximately 722.8 mm, which is lower than the state-wide average for Tamil Nadu. Figure 3 shows that the highest average annual rainfall in the basin, 1213.83 mm, occurs in Gopichettipalayam, while the lowest, 305.6 mm, is recorded in Punjapulipati. In 1981, the annual rainfall was at its minimum, recorded at 135.2 mm, compared to other years. As noted earlier, rainfall distribution across the basin is uneven, with central areas receiving more rainfall compared to other parts.

A unique feature of the lower Bhavani basin is the relatively high rainfall observed in its central region. Elevation plays a crucial role in shaping the rainfall distribution pattern, as highlighted by Geng et al. [10]. Spatial variability in the average annual rainfall is evident, with a coefficient of variation at 22.5%. This coefficient reflects the standard

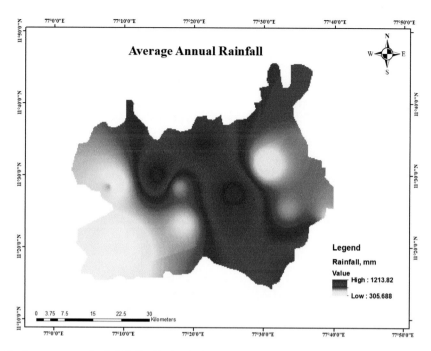

Fig. 3 Average annual rainfall distribution in lower Bhavani

deviation as a percentage of the mean annual rainfall, indicating the degree of variability across the basin (Table 2).

6.1 Temporal Variations in Annual Rainfall Distribution in Lower Bhavani Basin

The state Tamil Nadu has experienced huge variations in rainfall distribution for several years in the recent past due to the influence of climate change. The Bhavani river is one of the major sources of the Coimbatore water supply scheme. Therefore, the study of rainfall variability over the years is important and has a wide range of applications such as to know the local weather mechanisms [11] and the drastic changes in the ecosystem [12]. Therefore, in recent years, the study of rainfall variability in its spatial and temporal distribution has gained wide attention as it is the most important factor which determines the development and progress of the society and nation.

The rainfall was spatially averaged for the entire river basin using the IDW interpolation method. The minimum and maximum value of rainfall obtained in the basin in the years 1981 to 2019 was summarized in the Fig. 4. In the year 2000, the area has experienced the highest average rainfall of 1357.2 mm with minimum rainfall of 270 mm (near

Table 2 Average annual rainfall of different gauging stations

Sl. No.	Stations	Average annual rainfall, mm
1	Satyamanglam	1152.413
2	Athanai	418.7615
3	Bavnisgr_agr	419.4038
4	Anthiyur	657.4769
5	Gobichetipalayam	1214.243
6	Kavunthapadi	630.2846
7	Bavanisagar	626.0403
8	Chithode	502.4268
9	Kalingarayan	647.4111
10	Kongapalayam	1063
11	Lbpcanal33_7	967.0555
12	Lbpcanal54_1	869.6412
13	Nambiyur	536.4
14	Perundurai	1081.471
15	Punjpulipati	305.6105
16	Varattupalam	725.8714
17	Kodivri_acut	634.145
18	Annur	409.4407

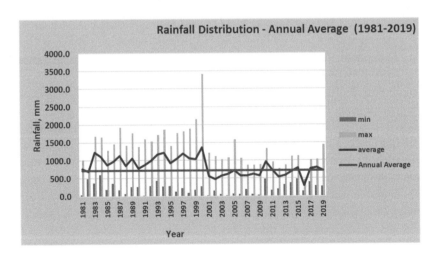

Fig. 4 Temporal distribution of annual rainfall in lower Bhavani basin (1981–2019)

to Punjapulipati) and maximum rainfall of 3422.6 mm (in LBPcanal 33_7). In this year, the area experienced average rainfall more than state's average of 900 mm.

Rainfall distribution was categorized into excess, normal, deficient, and scanty based on the following criteria: Excess rainfall is defined as 20% or more above the normal level, Normal ranges from +20% to −20% of the normal, Deficient refers to rainfall between −20% and −60% below normal, and Scanty represents a decrease of more than 60% below normal. This classification helps assess deviations from the typical rainfall pattern. In the Lower Bhavani River basin, there were no instances of scanty rainfall, but deficient rainfall occurred in the years 2001–2003, 2006–2007, 2009, and 2016. Conversely, the basin experienced excess rainfall during 1983–1987, 1989, 1991–2000, and 2010. Overall, the basin mostly experienced excess rainfall throughout the years, especially in the decade 1991–2000.

6.2 Seasonal Distribution of Rainfall

Rainfall patterns in the basin fluctuate significantly across different seasons, making the analysis of seasonal rainfall distribution crucial, particularly for the agricultural sector. Understanding these seasonal variations is essential for planning cropping cycles and ensuring sufficient irrigation to meet agricultural needs. As previously discussed, the main seasons in the region include winter, summer, SWM, and NEM.

Each of these seasons plays a vital role in shaping the region's agricultural practices. Winter rainfall, though generally lower, can be critical for certain crops, while summer rains help prepare the soil for the monsoon seasons. The SWM and NEM are the primary contributors to the overall water supply, replenishing reservoirs and groundwater essential for both irrigation and other water needs. Variations in rainfall across these seasons directly impact the timing and success of crop planting and harvesting cycles.

Figures 5a–d and 6 illustrate how rainfall is distributed seasonally across the Lower Bhavani River basin. By analysing these patterns, farmers can better plan agricultural activities, aligning them with the expected water availability during each season, thereby optimizing yields and managing water resources more effectively.

The SWM contributed average rainfall of 227.22 mm (31%), NEM which contributed 349.8 mm (47.8%) while Summer contributed 142.6 mm (19.5%) and Winter contributed 12.2 mm (1.7%) (Fig. 6).

The spatial variation in rainfall distribution across the Lower Bhavani basin during different seasons is depicted in Fig. 5a–d. The annual rainfall variability within the basin remains relatively consistent, with the coefficient of variation ranging between 21.7 and 33%. The highest variation (33%) occurs in the winter season. During the North-East monsoon, the variation is recorded at 29.8%, while the South-West monsoon shows a variation of 22.75%. The summer season experiences the least variability, with a coefficient of 21.7%.

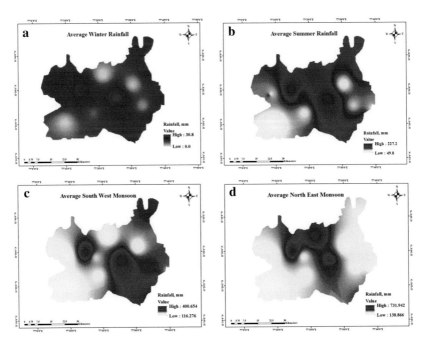

Fig. 5 **a**. Spatial distribution of average winter rainfall over lower Bhavani, **b**. Spatial distribution of average summer rainfall over lower Bhavani, **c**. Spatial distribution of average SWM rainfall over lower Bhavani, **d**. Spatial distribution of average NEM rainfall over lower Bhavani

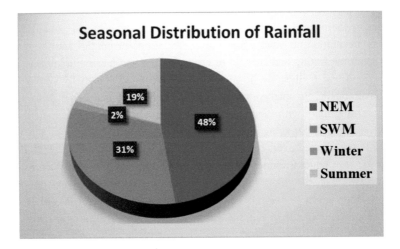

Fig. 6 Percentage contribution of rainfall in lower Bhavani basin- season wise

6 Results and Discussion

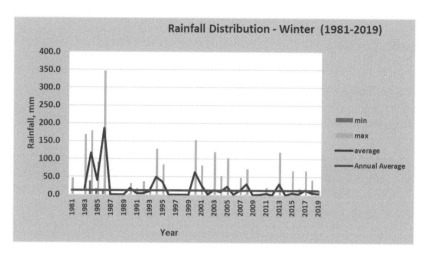

Fig. 7 Temporal distribution of winter rainfall in lower Bhavani basin (1981–2019)

Figure 7 illustrates the temporal variability of winter rainfall distribution in the Lower Bhavani basin from 1981 to 2019. Analysing this variability provides perceptions into long-term trends or patterns in rainfall over the area during this period. There was no recorded winter rainfall in the basin during the years 1987 to 1989, 1996 to 1999, 2009, 2012, and 2014. The highest average winter rainfall occurred in 1986, reaching 185.7 mm. Winter rainfall was significantly above long-term average in the years 1983–1986, 1990, 1991, 1994–1995, 2000–2001, 2005, and 2008. The basin experienced normal rainfall in 1981, 2003, 2007, and 2017. In the remaining years, the area received below-average or minimal rainfall. On an average, most gauging stations recorded less rainfall than the normal value of 12.2 mm during the winter season, with the highest rainfall distribution at the Gopichettipalayam gauging station, which recorded 30.8 mm on an average. This station received its maximum winter rainfall of 175.6 mm in 1984.

The average summer rainfall distribution in the Lower Bhavani basin ranges from 49.8 mm in the Punjapulipetti region to 227.2 mm near the Gopichettipalayam region, with a spatial average of 142.6 mm (Fig. 8). Analyzing the 39-year rainfall pattern reveals that six stations recorded normal rainfall, another six experienced deficient rainfall, five recorded excess rainfall, and one station had scanty rainfall.

Over the years, most stations received either normal or excess rainfall during the summer season. Figure 8 presents the temporal variation in summer rainfall distribution from 1981 to 2019. The years with excess rainfall include 1981, 1983–1984, 1986, 1988–1990, 1994, 1996–1997, 1999, 2004, 2009, 2011, 2015, and 2017–2018. Normal rainfall was observed in 1982, 1987, 1995, 2000–2001, 2005, 2008, 2010, 2012, and 2019. Conversely, deficient rainfall occurred in 1985, 1991–1993, 1998, 2002, and 2013–2014, while scanty rainfall was recorded in 2007 and 2016.

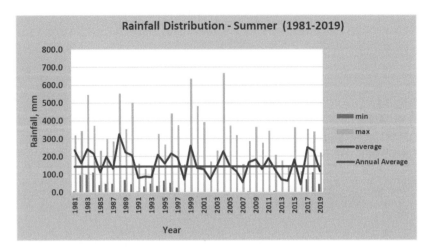

Fig. 8 Temporal distribution of summer rainfall in lower Bhavani basin (1981–2019)

The average rainfall distribution during the South-West monsoon season in the Lower Bhavani basin ranges from 116.4 mm in Punjpulipetti to 400.6 mm in Gopichettipalayam, with a spatial average of 227.2 mm. The highest recorded rainfall for this season was 929.6 mm, observed in the year 2000 at the Gopichettipalayam rain gauge station.

Figure 9 displays the temporal variation in rainfall distribution for the South-West monsoon from 1981 to 2019, highlighting significant fluctuations over the years. Notably, the basin experienced excess rainfall in several years, including 1981, 1983–1985, 1987–1989, 1991–1993, 1995–1998, 2000, 2014, and 2017–2018. These periods indicate a trend of above-average monsoon activity that likely contributed to better water availability in the basin.

Normal rainfall was recorded in the years 1986, 1994, 2001, 2007–2008, 2010–2011, 2013, 2015–2016, and 2019. These years reflect typical monsoon patterns, suggesting stable climatic conditions without extreme variations. Conversely, deficient rainfall was observed in all remaining years, indicating periods of below-average monsoon activity, which could have had negative impacts on agriculture and water resources in the region.

Interestingly, no instances of scanty rainfall were reported during the South-West monsoon season throughout the study period. This suggests that while there were years of deficit, the monsoon still brought significant precipitation, preventing extremely low rainfall conditions. Overall, this analysis indicates the variability of the monsoon and its critical role in the region's hydrological and agricultural cycles.

The average rainfall distribution during the North-East Monsoon (NEM) season in the Lower Bhavani basin varies significantly, ranging from 138.8 mm in Punjapulipatti to 731.9 mm in Kongapalayam, with a spatial average of 349.8 mm. The highest recorded

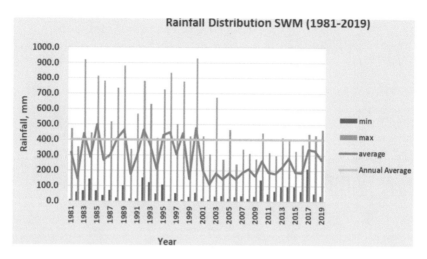

Fig. 9 Temporal distribution of SWM in lower Bhavani basin (1981–2019)

rainfall during this season was an exceptional 2339 mm, observed in the year 2000 at the LBP Canal_54_1 rain gauge station.

Figure 10 illustrates the temporal variation in rainfall distribution for the North-East Monsoon from 1981 to 2019. Notably, scanty rainfall was observed only in 1988 and 2016, indicating that extremely low rainfall during this season was relatively rare. Normal rainfall was recorded in the years 1982, 1986, 1989–1990, 1996, 2002, 2005–2007, 2011, 2018, and 2019. Excess rainfall was noted in several years, including 1983–1984, 1987, 1991–1994, 1997–2000, 2010, 2014, and 2015. These periods of abundant rainfall likely had a positive impact on water resources and agricultural yields in the basin. In contrast, deficient rainfall was recorded in the remaining years, indicating periods of reduced monsoon activity, which could have posed challenges for water availability and crop production.

6.3 Spatial Analysis of Seasonal Rainfall Distribution

- **Summer**: The highest rainfall is concentrated in the central and upper north-eastern parts, as well as the lower south-eastern regions of the basin. This distribution suggests localized convective systems or orographic effects influencing the rainfall pattern during this season.
- **South-West Monsoon (SWM)**: Maximum rainfall occurs in the south-eastern portion of the basin. This pattern could be attributed to the south-western winds bringing moisture from the Arabian Sea, leading to orographic lifting and enhanced precipitation in this area.

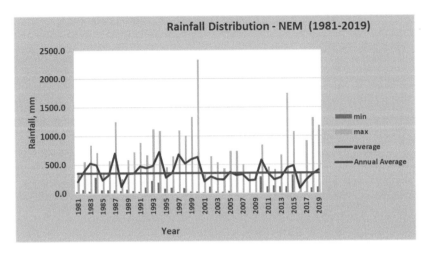

Fig. 10 Temporal distribution of NEM in lower Bhavani basin (1981–2019)

- **North-East Monsoon (NEM)**: The upper central region of the basin experiences the highest rainfall during this season.
- **Winter**: Rainfall is predominantly concentrated in the central part of the basin, possibly due to localized weather systems and cyclonic disturbances affecting the region during this season.

Overall, the spatial and temporal analysis of rainfall patterns in the Lower Bhavani basin highlights the complex interplay between monsoonal winds, topography, and local weather systems, which together shape the basin's hydrological characteristics.

6.4 Decade Wise Rainfall Distribution

Understanding temporal variations in rainfall distribution is crucial for planning and implementing effective land use and infrastructure developments. With this in mind, rainfall distribution data over the last four decades was analysed spatially. Figure 11a–d depicts the annual average rainfall distribution for different decades, covering the period from 1981 to 2019. The average annual rainfall recorded was 921 mm for the first decade (1981–1990), 936 mm for the second decade (1991–2000), 614.6 mm for the third decade (2001–2010), and 643 mm for the fourth decade (2011–2019).

The data indicates that the first two decades had similar rainfall patterns, but there was a significant 29% decline in the average annual rainfall from the first to the fourth decade. This reduction could be attributed to the effects of climate change, which may have altered regional weather patterns. Additionally, the variability in rainfall across the basin was

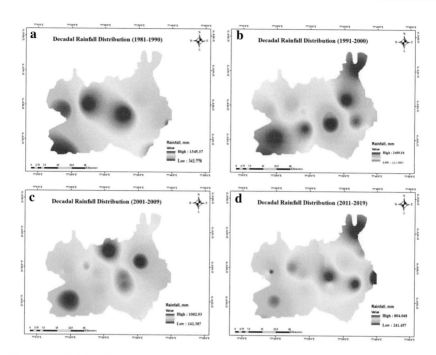

Fig. 11 a. Decadal distribution of average annual rainfall (1981–1990), b. Decadal distribution of average annual rainfall (1991–2000), c. Decadal distribution of average annual rainfall (2001–2010), d. Decadal distribution of average annual rainfall (2011–2019)

highest during the second decade, with a coefficient of variation of 39%, followed by the third decade at 25.4%. The spatial variability of average annual rainfall was relatively lower during the first decade (21.4%) and showed a further decrease in the last decade (10.3%). This trend suggests a narrowing range of rainfall distribution within the basin, indicating a shift towards more uniform, though reduced, rainfall across the region in recent years.

The maximum winter rainfall was recorded in the first decade (1981–1990) but showed a decreasing trend towards the last decade (2010–2019). The coefficient of variation (CV) within the basin increased significantly from 24.9% in the first decade to 41.6% in the last decade. This increase in variability suggests that the winter rainfall became less predictable over time, possibly due to changing weather patterns and reduced frequency of winter precipitation events (Fig. 12).

Summer rainfall distribution also exhibited notable changes over the decades, as shown in Fig. 13. The highest average summer rainfall was observed in the first decade (1981–1990), followed by a steady decline in the second (1991–2000) and fourth decades (2011–2019), with the lowest recorded in the third decade (2001–2010). The coefficient of variation was highest in the second decade (35.4%), indicating considerable spatial

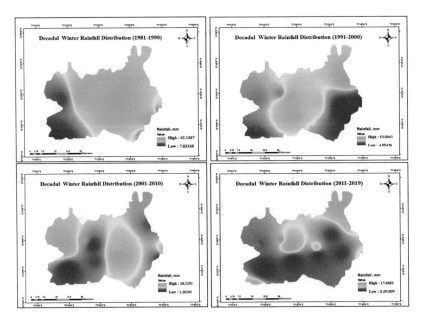

Fig. 12 Decadal distribution of winter rainfall

variability in rainfall distribution across the basin during this period. By the last decade, the CV had reduced to 12.2%, suggesting a more uniform but generally lower distribution of summer rainfall.

As depicted in Fig. 14, the SWM season showed the highest decadal average rainfall in the first (334.3 mm) and second decades (282.9 mm), followed by a significant decrease in the third (162.8 mm) and a slight recovery in the fourth decade (241.3 mm). The coefficient of variation was most pronounced in the second decade at 35.3%, reflecting substantial spatial differences in rainfall distribution within the basin. This variability declined to 25.4% in the third decade, indicating a shift towards more uniform but reduced monsoon rainfall, possibly affecting water availability and agriculture in the basin.

Figure 15 reveals that the NEM season exhibited the most consistent rainfall across the decades, with a peak average rainfall of 497.26 mm in the second decade. The average rainfall in the first, third, and fourth decades remained relatively stable at 315.1 mm, 308.1 mm, and 306.49 mm, respectively. The coefficient of variation, however, was consistently high across all decades, indicating considerable spatial variability in rainfall distribution. The CV values were 30% in the first decade, peaked at 39% in the second decade, and slightly decreased to 31% and 28% in the third and fourth decades, respectively. This high variability suggests that the NEM rainfall is highly influenced by localized factors such as topography and microclimatic conditions, making it less predictable and more challenging for water management and agricultural planning.

6 Results and Discussion

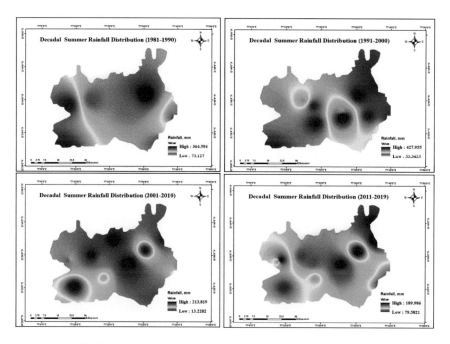

Fig. 13 Decadal distribution of summer rainfall

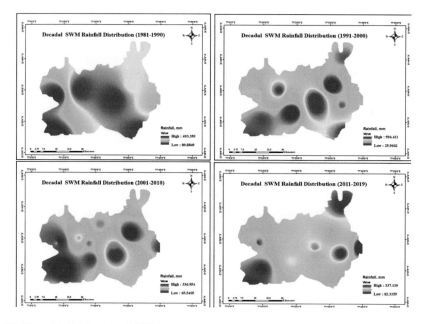

Fig. 14 Decadal distribution of SWM

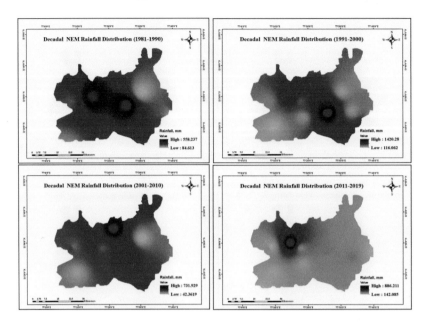

Fig. 15 Decadal distribution of NEM rainfall

The analysis highlights significant temporal and spatial variability in rainfall across the basin, with distinct changes in each season over the decades. The decreasing trend in average annual rainfall, especially during the winter and summer seasons, and the fluctuating patterns during the monsoon seasons, point to potential impacts of climate change. These variations in rainfall distribution pose challenges for sustainable water resource management and agricultural productivity in the region. Effective adaptation strategies, such as improved water storage infrastructure, better irrigation practices, and diversified cropping patterns, are essential to mitigate the adverse effects of these changes and ensure the resilience of the basin's agricultural and water systems.

The seasonal distribution of rainfall in the Lower Bhavani basin has undergone significant changes over the four decades from 1981 to 2019. Analysing the proportion of total annual rainfall contributed by each season provides insights into shifting rainfall patterns and their implications for water resources and agriculture. Below is a detailed analysis based on the given percentages for winter, summer, South-West Monsoon (SWM), and North-East Monsoon (NEM) seasons.

During the first decade, the SWM season was the dominant contributor, accounting for 36.4% of the annual rainfall, followed closely by the NEM at 34.3%. This balanced distribution between the two monsoon seasons suggests that the basin had a relatively stable water supply throughout the year. The summer season contributed a significant

6 Results and Discussion

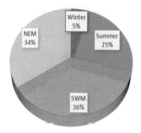
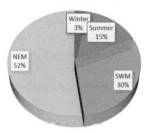

Fig. 16 Percentage contribution seasonal distribution of rainfall (in first two decades)

24.9%, indicating active pre-monsoon weather systems. Winter rainfall, though minimal at 4.4%, still played a minor role in the total rainfall distribution (Fig. 16).

In the second decade, the NEM season became the predominant source of rainfall, contributing 52.1% of the annual total, more than half of the yearly rainfall. This indicates a shift towards increased reliance on the NEM for water resources, possibly due to a decrease in SWM contribution, which dropped to 29.7%. The summer season's share reduced significantly to 15.4%, and winter rainfall further decreased to 2.8%, highlighting a decline in rainfall outside the main monsoon seasons. This shift could indicate a potential vulnerability to water shortages during the non-monsoon months, impacting agriculture and water management (Fig. 16).

The third decade continued the trend of high reliance on the NEM season, which contributed 50% of the total annual rainfall. The SWM contribution further declined to 26.4%, while the summer season's share increased to 21.5%, indicating a resurgence of pre-monsoon rainfall activities. Winter rainfall further declined to just 2%, showing its diminishing role in the overall water availability. This decade's pattern reflects a growing disparity between the two monsoon seasons, with the NEM becoming increasingly crucial for meeting water demands (Fig. 17).

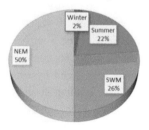
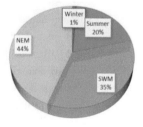

Fig. 17 Percentage contribution seasonal distribution of rainfall (in last two decades)

In the final decade, the NEM season's contribution decreased slightly to 44.2%, while the SWM contribution recovered to 34.8%. This indicates a slight rebalancing between the two monsoon seasons, though the NEM still dominates. The summer season continued to contribute a significant share at 20.1%, suggesting continued pre-monsoon rainfall activities. Winter rainfall dropped to a negligible 1%, highlighting its almost complete disappearance from the seasonal distribution (Fig. 17).

The analysis of decadal rainfall patterns in the Lower Bhavani basin highlights significant shifts, with the North-East Monsoon (NEM) emerging as the primary contributor to annual rainfall, especially in the second and third decades, where it accounted for over 50% of the total. This increasing reliance on the NEM, likely influenced by broader climatic changes, makes the region more vulnerable to any variability in this season's rainfall. Concurrently, winter rainfall has seen a consistent decline from 4.4% in the first decade to just 1% in the last, indicating a substantial reduction in winter precipitation that could lead to prolonged dry spells and increased pressure on the monsoon seasons to meet water demands. The summer season also exhibited fluctuations, with a significant decline in the second decade followed by partial recovery in the subsequent decades, reflecting its inconsistent role as a supplemental water source and complicating water management efforts. Meanwhile, the South-West Monsoon (SWM) showed a slight recovery in the last decade, with its contribution rising to 34.8%, up from 26.4% in the third decade, yet still below the first decade's 36.4%. This partial rebound suggests some stabilization in SWM patterns, but overall, the shifting dynamics across all seasons underscore the need for adaptive water management and agricultural strategies to address the growing variability and uncertainty in rainfall distribution.

7 Conclusion

The detailed analysis of rainfall distribution in the Lower Bhavani river basin reveals significant temporal and spatial variability over the past four decades, with distinct shifts in seasonal patterns. The decadal trend indicates an increasing dependence on the North-East Monsoon (NEM), which has become the primary contributor to annual rainfall, particularly in the last two decades, making the basin more susceptible to fluctuations in this season. Concurrently, the steady decline in winter rainfall, coupled with the inconsistent performance of the summer season, poses additional challenges for water resource management and agricultural sustainability. The partial recovery of the South-West Monsoon (SWM) in the last decade offers a glimmer of stability; however, its contribution remains below earlier levels. These changing rainfall dynamics, influenced by broader climatic changes, highlight the urgent need for adaptive water management strategies, including enhanced water storage infrastructure, efficient irrigation practices, and diversified cropping patterns, to mitigate the adverse impacts on agricultural productivity and ensure the long-term resilience of the basin's water resources.

References

1. Gonçalves, H. C., Mercante, M. A., & Santos, E. T. (2011). Hydrological cycle. *Brazilian Journal of Biology, 71*, 241–253.
2. Batisani, N., & Yarnal, B. (2010). Rainfall variability and trends in semi-arid Botswana: implications for climate change adaptation policy. *Applied Geography, 30*(4), 483–489.
3. Ayanlade, A., Radeny, M., Morton, J. F., & Muchaba, T. (2018). Rainfall variability and drought characteristics in two agro-climatic zones: An assessment of climate change challenges in Africa. *Science of the Total Environment, 630*, 728–737.
4. Strauch, A. M., MacKenzie, R. A., Giardina, C. P., & Bruland, G. L. (2015). Climate driven changes to rainfall and streamflow patterns in a model tropical island hydrological system. *Journal of Hydrology, 523*, 160–169.
5. Jan, C. D., Chen, T. H., & Lo, W. C. (2007). Effect of rainfall intensity and distribution on groundwater level fluctuations. *Journal of hydrology, 332*(3–4), 348–360.
6. Wasko, C., & Nathan, R. (2019). Influence of changes in rainfall and soil moisture on trends in flooding. *Journal of Hydrology, 575*, 432–441.
7. CGWB [Central Ground Water Board], Aquifer Mapping and Ground Water Management, 2017.
8. Lee, J., Kim, S., & Jun, H. (2018). A study of the influence of the spatial distribution of rain gauge networks on areal average rainfall calculation. Water (Switzerland), 10(11). https://doi.org/10.3390/w10111635.
9. Yang, X., Xie, X., Liu, D. L., Ji, F., & Wang, L. (2015). Spatial Interpolation of Daily Rainfall Data for Local Climate Impact Assessment over Greater Sydney Region. Advances in Meteorology, 2015.
10. Geng, H., Pan, B., Huang, B., Cao, B., & Gao, H. (2017). The spatial distribution of precipitation and topography in the Qilian Shan Mountains, northeastern Tibetan Plateau. Geomorphology, 297, 43–54.
11. Joshi, S., Kumar, K., Joshi, V., & Pande, B. (2014). Rainfall variability and indices of extreme rainfall-analysis and perception study for two stations over Central Himalaya, India. Natural Hazards, 72(2), 361–374.
12. Sillmann, J., Thorarinsdottir, T., Keenlyside, N., Schaller, N., Alexander, L. V., Hegerl, G., … Zwiers, F. W. (2017). Understanding, modeling and predicting weather and climate extremes: Challenges and opportunities. Weather and Climate Extremes, 18(October), 65–74.

Geomorphometric Analysis of Nileswar Sub-Watershed, Kerala Using GIS and Remote Sensing

Abstract

The Nileswar sub-watershed in the Kasaragod district, located in Northern Kerala, underwent an analysis as part of planning watershed management activities. Geospatial tools, including remote sensing and GIS, have been employed to delineate watershed boundaries and derive stream networks. The Shuttle Radar Topographic Mission (SRTM) Digital Elevation Model (DEM) data was utilized for morphometric analysis, which involved the evaluation of various morphometric parameters like linear characteristics, areal properties, and relief features. These morphometric parameters were examined using established methods such as those pioneered by Horton and Strahler. To facilitate this, micro-watersheds were defined using the DEM data using ArcGIS software. An array of watershed parameters, including stream order, stream length, stream frequency, drainage density, slope, bifurcation ratio, form factor, circulatory ratio, elongation ratio, and infiltration number, were computed individually for each of the nine micro-watersheds through geospatial techniques. The sub-watershed was identified as fifth order stream. The compound scores were calculated from these morphometric parameters based on the compound ranking procedure. Based on their prioritized scores, the micro-watersheds were categorized into three classes: high (<4.7), medium (4.7–5.3), and low (>5.3) priority. In this study, micro-watershed 6 has been identified as a high-priority area requiring urgent soil conservation measures.

Keywords

Elongation ratio · Morphology · Nileswar sub-watershed · Relief aspect

1 Introduction

Land and water are the two fundamental natural resources in the world. Due to the strong interaction between these two resources, the characteristics of land and water varies spatially as well as temporarily. Extreme climatic events, such as heavy rainfall, exert a significant impact on the geospatial attributes of watersheds. These events result in adverse consequences for both land and water resources, including sediment accumulation, alterations in mean annual water flow, increased peak discharges, flooding, and the contamination of water and soil. Remarkably, many watersheds lack comprehensive strategic plans to effectively address and mitigate these issues. To conduct a thorough risk assessment of soil erosion, floods, runoff, drought, and related challenges, it is imperative to possess an intricate understanding of the physical properties and dynamics of the watershed [1].

The morphometric characteristics of a watershed play a crucial role in shaping its sustainability in agriculture. Understanding the topographic features of a watershed, including aspects like drainage density, slope, and form factor, is essential for efficient land use planning and natural resource management. These characteristics influence the flow of water, sediment transport, and soil erosion patterns within the watershed. Sustainable agriculture practices heavily depend on the ability to manage water resources effectively, control erosion, and optimize land use. A watershed with favourable morphometric attributes, such as low slope and high drainage density, can enhance water availability and reduce the risk of soil erosion. By considering these morphometric features, farmers and land managers can make informed decisions on crop selection, irrigation strategies, and soil conservation practices, ultimately promoting long-term sustainability in agriculture within the watershed.

Morphometric analysis pertains to the examination and assessment of the Earth's topography, which comprises both on-site descriptions of landforms and the quantitative mathematical analysis of surface features., shape as well as dimensions and morphological mapping. The term geo-morphometry was derived from the Greek words and morphometric studies in the hydrology field was first started by Horton [2]. This analysis is crucial for understanding the hydrological processes, sediment transport, and erosion patterns within a watershed, which are essential for effective water resource management and conservation efforts.

Morphometric analysis of a watershed or drainage basin typically comprises three primary facets: linear characteristics, areal properties, and relief features. Each parameter under these aspects is defined using mathematical equations. These morphometric parameters serve as key indicators of the hydrological behavior of the basin, demonstrating a strong correlation with the impact of precipitation and streamflow within the watershed [3].

Nowadays, studying the hydro-geomorphological concept about the watershed is very easy to understand because of the introduction of modern tools and software for terrain analysis and surface modelling [4, 5] Geographic Information Systems (GIS) and remote sensing technologies have revolutionized the field of watershed management by providing efficient tools for morphometric analysis. GIS facilitates the integration, storage, analysis, and visualization of spatial data related to a watershed. Remote sensing techniques, on the other hand, enable the acquisition of high-resolution imagery and elevation data covering large areas, which are vital inputs for morphometric analysis. The synergy between GIS and remote sensing allows for detailed assessment and characterization of watersheds.

The present study aims to implement GIS and remote sensing techniques in morphometric analysis of the Nileswar sub-watershed and assess the spatial distribution of morphometric parameters for effective watershed management planning. Additionally, it aims to understand the hydrological response of the sub-watershed.

2 Materials and Methods

2.1 Description of the Study Area

The Nileswar sub-watershed was selected for the study. The location map of the selected sub-watershed is shown in Fig. 1. The sub-watershed is located at the northern part of Kerala. The total catchment area of the river is about 190 km^2. The area is located in between the latitude–longitude range of 12°13′N, 75°05′E and 12°23′N, 75°17′E. The sub basin lies in 43 zone in UTM projection. This sub-watershed lies in Kasaragod district of Kerala. The Nileswar sub-watershed is situated in a humid tropical climate region and experiences an annual average rainfall of approximately 3600 mm. Rainfall in this area exhibits seasonal variation, with the majority of precipitation occurring during the South-West monsoon season. The sub-watershed also displays spatial diversity in terms of land use patterns and soil characteristics within its catchment area.

The Nileswar river, locally referred to as Thejaswini puzha, originates from the Kinanur hills in the Kasaragod district of Kerala. It meanders through the Vellarikundu and Hosdurg taluks, comprising various localities such as Balal, Kodom-Belur, Kinanoor Karindalam, and Madikai, as well as passing through the municipalities of Kanhangad and Nileshwar. Its journey culminates at 'Oorcha' in Nileshwar, where it merges with the Kariangode river before flowing into the Arabian sea. The Nileswar river receives contributions from several tributaries, including the Edathod river, Mayanganam river, Aryangal thodu, and baigot hole.

2.2 Methodology

For the geo-morphometric analysis of the study area, data from the Shuttle Radar Topographic Mission (SRTM) digital elevation model (DEM) with a resolution of 30 m were

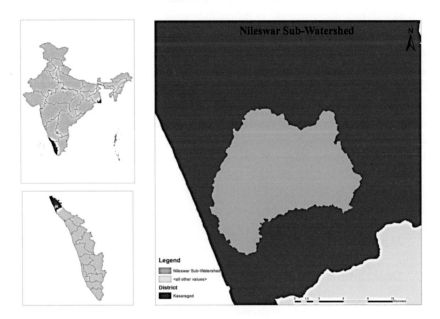

Fig. 1 Location of the study area

utilized. These DEM data were acquired from the USGS Earth Explorer. The analysis was conducted within a GIS framework, specifically employing ArcGIS 10.8 software.

The morphometric analysis of the watershed geometry encompassed the computation of three key aspects:

1. **Linear characteristics of drainage network**: This involves the measurement and assessment of the linear features within the drainage network.
2. **Areal properties of basin**: It pertains to the quantitative examination of the entire basin area.
3. **Relief features of drainage channel**: This aspect focuses on the evaluation of the elevation differences within the drainage channels.

Various mathematical relationships and methodologies were employed to derive a range of watershed parameters, which are detailed in the following section.

2.2.1 Delineation of Sub-Basin Boundary

The Hydrology tool within ArcGIS software can be employed to digitally demarcate the watershed boundary based on the Digital Elevation Model (DEM) of the study area. The same can also be easily performed using the ArcHydro plugin. The delineation process using the hydrology tool involves several key steps, which include the creation of the following raster layers.

1. Fill sink raster
2. Flow direction layer
3. Flow accumulation layer
4. Watershed raster

The resultant watershed rasters are subsequently converted into polygon features through conversion tools. From this pool of polygon features, a specific polygon corresponding to the study location is chosen. Moreover, stream networks are extracted from the flow accumulation raster using appropriate threshold values.

2.2.2 Attribute Data Management in ArcGIS

In this study, most of the parameters were computed using the Attribute data management in ArcGIS, which refers to the process of organizing, editing, and managing the non-spatial data associated with geographic features in a GIS software. In ArcGIS, attribute data is typically stored in attribute tables, which are database-like tables associated with geographic features. These tables contain information about the attributes, characteristics, or properties of spatial features, and they play a crucial role in GIS analysis, visualization, and decision-making. For example, in the attribute table corresponding to any drainage network map, it is customizable to add different stream features such as length, stream ID, stream order, water quality, etc.

Here's an overview of attribute data management in ArcGIS:

1. **Attribute tables**: Each feature class or layer in ArcGIS has an associated attribute table. These tables store information as rows and columns, where each row represents a feature, and each column represents an attribute or field. Fields can contain various types of data, such as text, numbers, dates, and more.
2. **Data entry and editing**: Attribute data can be added or edited directly within ArcGIS using the attribute table. You can create new records, modify existing data, and delete records as needed. This is useful for updating feature attributes.
3. **Data validation**: ArcGIS provides tools for data validation to ensure that attribute data adheres to specified rules and constraints. This helps maintain data accuracy and consistency.
4. **Relational databases**: ArcGIS can work with relational databases (e.g., SQL Server, Oracle, and PostgreSQL) to manage attribute data more efficiently. This allows for more complex data management and querying.
5. **Joining and relating tables**: You can join attribute tables to combine data from different sources based on common fields, or you can establish relationships between tables to maintain the integrity of related data.
6. **Calculating fields**: ArcGIS allows you to perform calculations on attribute data using the Field Calculator. This can be used to create new fields, derive values from existing ones, or perform various data transformations.

7. **Data querying and analysis**: You can use attribute data to perform queries and analysis, filtering features based on their attributes. This is valuable for generating reports, thematic mapping, and decision support.
8. **Metadata management**: You can document and manage metadata for attribute data, which is essential for data sharing and understanding the context of the information.
9. **Data export and import**: ArcGIS supports various data formats, allowing you to export attribute data to spreadsheets, databases, or other GIS software for further analysis. You can also import external attribute data into your GIS project.

Proper attribute data management is crucial for maintaining the accuracy and integrity of GIS data, and it enhances the utility of a GIS for decision-making, spatial analysis, and visualization. It ensures that the non-spatial information associated with geographic features is well-organized and readily accessible. Hence, this feature is considered a valuable asset for determining morphometric parameters of any watershed.

2.2.3 Linear Characteristics of Drainage Network

The linear characteristics of the drainage network include stream order, stream count, bifurcation ratio, the length of individual streams, ratio of stream lengths, and the average stream length.

Stream order is a vital parameter for characterizing watershed geometry, offering insights into the hierarchical structure of the stream networks. stream order maps are a valuable resource for designers and watershed managers, helping them make proper decisions about the location and type of groundwater recharge structures like check dams and gabions. By considering stream order and other relevant factors, they can enhance water resource management and environmental conservation actions within the watershed.

The Strahler method, as outlined by Strahler [6] is applied to determine stream order. In this method, the smallest streams within the basin hold the lowest stream order, while the largest stream in the system carries the highest stream order. When two or more streams of a specific order combine, they give rise to a stream of the next higher order. In the context of this study, stream order delineation was accomplished using the stream order tool in ArcGIS 10.8. This process involved providing both the stream raster and flow direction raster as input data [7].

In this study, we ascertained the stream number (N_u), which denotes the total count of streams with a specific stream order, using attribute data management techniques within the GIS environment.

The bifurcation ratio (R_b) is defined as the ratio of the number of streams (N_u) within a specific order 'u' to the number of streams (N_{u+1}) in the next higher order. This relationship can be expressed using the following formula (Eq. 1).

$$R_b = \frac{N_u}{N_{u+1}} \tag{1}$$

2 Materials and Methods

Another crucial parameter under consideration in this study is Stream Length (L_u), which signifies the length of streams within each particular stream order 'u'. This parameter holds great importance in morphometric analysis because it is affected by both the slope of the terrain and the soil characteristics of the area. In general, smaller stream lengths tend to indicate regions with steeper topography and finer-textured soils, as outlined by Sreedevi et al. [8]. To determine L_u in this study, the calculation was conducted using attribute data management techniques, specifically employing the "calculate geometry" option within ArcGIS software.

Using the stream length (L_u) and stream number (N_u) data, the mean stream length was calculated. Mean Stream Length is determined as the ratio of the total length of streams within a specific order to the number of streams of that same order [9].

Another linear aspect addressed in this study is Stream Length Ratio (Lr), which is defined as the ratio between the average lengths of streams within a particular order and the average lengths of streams in the next lower order.

2.2.4 Areal Properties of Drainage Basin

It includes a range of parameters, including the area, length, width, and perimeter of the basin, as well as metrics like the lemniscate, form factor, circulatory ratio, elongation ratio, Rho coefficient, stream density, stream frequency, drainage density, drainage texture, coefficient of channel maintenance, and infiltration number. These parameters help us understand the shape, drainage patterns, and physiographic development of the region. Each parameter serves a unique purpose in the analysis:

Basin area and perimeter are two significant watershed parameters. In this study, the area (A) and perimeter (P) of the watershed basin were determined using the "calculate geometry" tool within ArcGIS 10.8. Basin length (L_b) is characterized as the measurement of length extending from the basin outlet to the furthest point along the boundary of the basin [9]. Whereas basin width is described as the ratio of the basin area to the basin length (L_b), which can be expressed mathematically as follows (Eq. 2).

$$w = \frac{A}{L_b} \qquad (2)$$

Another morphometric parameter of interest is the Lemniscate (K), a measure used to determine the slope of the watershed. This parameter, as defined by Chorley [10], plays a crucial role in understanding the terrain characteristics of the watershed. Mathematically the Lemniscate (K) can be expressed as follows:

$$K = \frac{L_b^2}{4 \times A} \qquad (3)$$

A larger K value generally suggests a flatter terrain, while a smaller K value indicates a steeper slope within the watershed.

Form Factor (R_f) is another morphometric parameter introduced by Horton [11] represents the ratio of the basin area to the square of the basin length. R_f provides insights into the compactness or elongation of the watershed.

$$R_f = \frac{A}{L_b^2} \qquad (4)$$

Circulatory Ratio (R_c) first conceptualized by Miller [12], this metric defines the shape of the basin. It is the ratio of the basin area (A) to the area of a circle with the same perimeter as the basin. It is a dimensionless value. A higher Rc suggests a more centralized flow, while a lower value indicates a potentially meandering or irregular flow pattern.

The Elongation Ratio (R_l) is another morphometric parameter used to assess the shape of a watershed or drainage basin. It was introduced by Schumm [13]. The primary purpose of the R_l is to quantify and characterize the degree to which a watershed is elongated or stretched in shape. Mathematically, it is defined as the ratio between two essential dimensions: the diameter of a hypothetical circle (D) with an area equivalent to that of the watershed, and the maximum length of the basin (Lbm), which represents the longest distance between two points along the basin's boundary (Eq. 5).

$$R_l = \frac{D}{L_{bm}} \qquad (5)$$

The Elongation Ratio value ranges between 0 and 1, with typical values often falling within the range of 0.4–1.

The Rho coefficient is one of the important parameters of morphometry relating drainage density to physiographic development of a watershed which helps to evaluate the storage capacity of drainage networks [2]. It can be calculated as the ratio between stream length ratio and bifurcation ratio.

Stream Density (F) measures the total number of stream segments of all orders per unit basin area [11]. Mathematically, it's expressed as Eq. (6).

$$F = \frac{\sum N_u}{A} \qquad (6)$$

Drainage density (D_d) is defined as the stream length per unit area in a watershed region, as per Horton [2]. It quantitatively analyzes the landform. Drainage density (Dd) values are categorized into several classes: low (<1), moderate (1–2), high (2–3), and very high (3–4). The stream lengths were calculated using the calculate geometry option in the attribute table of ArcGIS 10.8. In addition, drainage texture (Dt), also discussed in this study and is defined as the ratio of the total number of stream segments of all orders to the basin perimeter [2]. It serves as a crucial parameter in geomorphological analysis, offering insights into the relative spacing of drainage lines. Based on D_t values,

2 Materials and Methods

the classification, as established by Smith in 1939, includes five categories: very coarse (<2), coarse (2–4), moderate (4–6), fine (6–8), and very fine (>8).

The inverse of the drainage density is known as constant of channel maintenance (C). It is expressed as km^2/km. The value increases as the area increases. The C values convey information about the susceptibility of soil to erosion, with different ranges indicating varying degrees of erodibility. These degrees include: <0.2 (Highly erodible), 0.2–0.3 (Mildly erodible), 0.3–0.4 (mildly less erodible), 0.4–0.5 (Slightly erodible), and >0.5 (very slightly erodible) [14].

Infiltration Number (In) is another parameter considered in this study, and is expressed as the result of multiplying drainage density by stream frequency, offering insights into the watershed's infiltration characteristics [15].

2.2.5 Relief Features of Sub-Watershed

This category mainly consists of slope, aspect, relief, maximum basin relief and relative relief. The calculation of each of these parameters were given in this section.

The study delves into the fundamental characteristics of the watershed, with a particular focus on slope and aspect, both of which play pivotal roles in shaping the overall landscape. The slope, a critical attribute, was calculated using the SRTM DEM and the spatial analyst tool within ArcGIS 10.8. Concurrently, aspect, which pertains to the direction of the slope surface, was derived from the same DEM dataset using ArcGIS's Spatial Analyst tool. The outcome is a detailed raster map that provides a comprehensive depiction of aspect, featuring values spanning from 0°, representing true north, to 360°, with 90° indicating an eastward orientation, 180° signifying the south, and so forth. These findings contribute valuable insights into the terrain's topographical makeup, aiding in a deeper understanding of the watershed's characteristics and orientation.

Maximum basin relief (H) is a critical topographic feature of the watershed, representing the elevation difference between the catchment outlet and the highest point along the watershed boundary [16, 17]. Additionally, the concept of *relative relief*, introduced by Melton in 1957, serves as an important measure for assessing the watershed's relief. Mathematically, relative relief can be calculated as the ratio of H (maximum basin relief) and P (basin perimeter).

2.2.6 Watershed Prioritization

The watershed prioritization was carried out using nine morphological parameters estimated for nine individual micro-watersheds (MWS) within the sub-watershed. The delineated micro watershed map is shown in the Fig. 2, comprising nine micro watersheds, are denoted as MWS 1, MWS 2 and so on. The priority ranking of micro watersheds for conservation and planning activities was carried out by compound ranking method and is summarizes in Tables 3 and 4 [18].

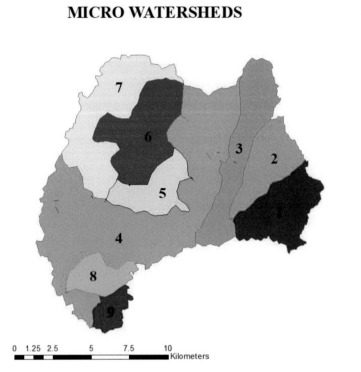

Fig. 2 Micro-watersheds of Nileswar sub-watershed

3 Results and Discussion

The study area's terrain characteristics, including the DEM, flow direction map, and stream network map (Figs. 3, 4 and 5), form the foundation of this analysis. The Nileswar sub-watershed's morphometric attributes have been quantified and are detailed in the subsequent section. These morphometric parameters were precisely determined using attribute data management techniques. The extraction of stream networks within the Nileswar sub-watershed was executed. Notably, the watershed exhibits a dendritic drainage pattern, characterized by a multitude of streams converging as tributaries into the river channel. This pattern is indicative of a terrain where the river closely follows the natural slope of the landscape, as elucidated by Hejran and Singh [9].

3 Results and Discussion

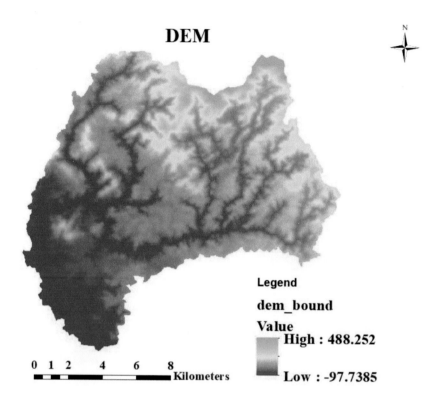

Fig. 3 DEM of Nileswar sub-watershed

3.1 Linear Aspects of Drainage Basin

The derived drainage map and stream order map of the Nileswar sub-watershed is given in Figs. 5 and 6. Stream order ranges from one to five. There are nine micro-watersheds, out of these six micro-watersheds has fifth ordered (MWS 1, 2, 3, 5, 6, and 8). MWS 4 has fourth ordered, MWS 7 has third ordered and MWS 9 has second ordered streams. Stream order maps are valuable tools for various applications in hydrology and watershed management, including the identification and design of groundwater recharge structures like check dams and gabion structures.

Higher-order streams (e.g., 5th order) are typically larger and have more significant flow, while lower-order streams (e.g., 1st order) are smaller and often temporary. This classification helps in understanding the flow patterns in a watershed. Designers can use these maps to identify areas within the watershed where they want to construct groundwater recharge structures. For instance, if the objective is to build cement concrete check dams, they might prefer sites with stream orders ranging from 2 to 4 [19].

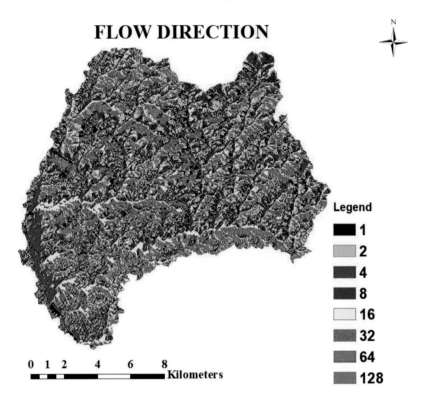

Fig. 4 Flow direction map of Nileswar sub-watershed

3.1.1 Stream Number (N_u)

The distribution and hierarchy of streams in a watershed are of paramount importance in watershed modeling. The number and arrangement of streams play a pivotal role in determining the flow of water, sediment transport, and the overall hydrological behavior of the area. Understanding the distribution of streams, such as the prevalence of first order streams over higher-order ones, is crucial for modeling. It helps in the accurate simulation of runoff, as first order streams often respond more rapidly to rainfall events. Meanwhile, the convergence of streams into higher order channels can influence flow dynamics and flood propagation.

In this sub-watershed's morphometric analysis, we observe a distinct hierarchical pattern in the distribution of streams originating from hilly terrains. First order streams, the smallest and most numerous, number 224 in this region, while second order streams, formed when two first order streams merge, total 62. The sub-basin contains streams of up to five hierarchical orders, amounting to 292 streams across all orders.

This hierarchy is a common characteristic of stream networks. Additionally, stream distribution varies across the sub-basin, with MWS 4 housing the highest number of streams

3 Results and Discussion

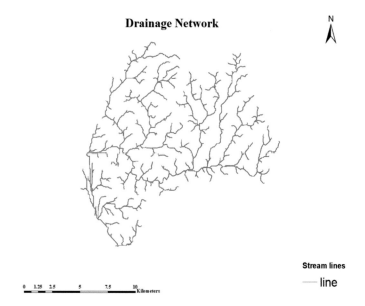

Fig. 5 Stream network of Nileswar sub-watershed

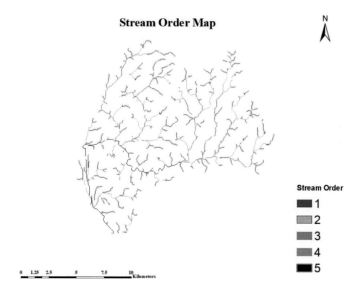

Fig. 6 Stream order map of Nileswar sub-watershed

(90) and MWS-9 having the fewest (eight). These distribution patterns are influenced by dynamic hydrological conditions and relief parameters within the basin, highlighting the importance of local environmental factors in shaping stream networks in this region. Understanding this hierarchy is essential for various hydrological and environmental studies.

3.1.2 Bifurcation Ratio (R_b)

The calculated bifurcation ratios for the Nileswar sub-watershed, corresponding to each stream order, are presented in Table 1. Notably, it is observed that the bifurcation ratio (R_b) does not remain constant when moving from one stream order to the next. These variations can be attributed to the geological and lithological characteristics of the catchment, as described by Strahler [6].

Geologically, the Nileswar sub-watershed is an integral part of the Western Ghat river system. The primary geological feature in the area is the presence of the Peninsular Gneissic Complexes (PGC) as the basement rocks. It's important to note that these rivers are not structurally controlled, meaning their courses are not primarily determined by geological fault lines or structural features. Instead, the topography of the terrain is primarily shaped by the formation of the Western Ghats, which influences the flow patterns. Additionally, the development of the South-West monsoon plays a pivotal role in influencing the hydrology of these watersheds.

The combination of these geological factors and the impact of the monsoon season contributes to the dynamic nature of the bifurcation ratios observed within the Nileswar sub-watershed. The irregularities in R_b can be linked to the complex interplay between geological features, topographic relief, and hydrological processes in this region.

The calculated bifurcation ratio for a Nileswar sub-watershed corresponding to each stream order was given in Table 1. It is observed from the R_b is not same from one order to its next order. These irregularities are dependent upon the geological and lithological development of the catchment [6]. The R_b is dimensionless property and the lower values of R_b represents the watersheds with less structural disturbances [6] and the drainage

Table 1 Linear aspects parameters for Nileswar sub-watershed

Stream order	Stream number	Stream length, km	Mean stream length	Stream length ratio	Bifurcation ratio
I	224	325.02	1.45		
II	62	157.53	2.54	1.75	3.61
III	24	87.42	3.64	1.43	2.58
IV	5	41.34	8.27	2.26	4.8
V	1	54.78	54.78	6.63	5
Total	292	666.09			

3 Results and Discussion

Table 2 Weighted bifurcation ratio

Stream order	Stream number	R_b	No. of streams used in the ratio (Nur)	$R_b * Nur$
I	224			
II	62	3.61	286	1032.46
III	24	2.58	86	221.88
IV	5	4.8	29	139.2
V	1	5	6	30
Total	292	15.99	407	1423.54
Weighted $R_b = 3.49$		Mean $R_b = 3.99$		

pattern has not been distorted because of the structural disturbances [20]. In the particular study, R_b value varies from 2.58 to 5 with mean bifurcation ratio of 3.99 and weighted R_b of 3.49. The estimated weighted R_b was given in Table 2. The variation in bifurcation ratio in each micro watershed is given in Table 3. The higher R_b was observed in the MWS-2, MWS-3 and MWS-9. The higher values of R_b indicates early hydrograph peak with a potential for flash flooding. So micro-watersheds with higher bifurcation ratios were given more priority [18].

3.1.3 Stream Length and Stream Length Ratio

The sub-watershed's stream length was accurately determined through attribute data management, a crucial metric with significant implications for the area's hydrological characteristics. Stream length plays a pivotal role in determining drainage density and the extent of land contributing to runoff within the watershed. In this context, the total cumulative length of all streams, comprising various hierarchical orders, amounts to 666.09 km. This value reflects the extensive network of watercourses within the sub-watershed.

The specific breakdown of these stream lengths reveals the distribution of the watercourses by their order. First-order streams encompass a total length of 325.02 km, while second-order streams extend to 157.53 km. This data, as detailed in Table 1, offers valuable insights into the hierarchical organization of the sub-watershed's drainage system.

To gain a deeper understanding of the sub-watershed's hydrology, mean stream length is a key parameter. It is calculated by dividing the length of streams within each order by the number of streams at that order. In the case of the Nileswar sub-watershed, mean stream length spans a considerable range, from 1.45 to 54.78 km, as presented in Table 1.

The significance of mean stream length is twofold. Firstly, it serves as a fundamental indicator of the stream network's spatial distribution. Smaller mean stream lengths suggest a denser network of smaller streams, which can have implications for flood risk, water quality, and habitat diversity. On the other hand, larger mean stream lengths indicate

Table 3 Result of morphometric analysis for nine micro-watersheds

Micro-watersheds	Bifurcation ratio (R_b)	Slope (%)	Stream frequency (F)	Drainage density (D_d)	Drainage texture (D_t)	Elongation ratio (R_l)	Infiltration Number (In)	Form factor (R_f)	Circulatory ratio (R_c)
MWS-1	3.78	15.00	1.50	3.35	1.19	0.71	5.03	0.46	0.45
MWS-2	7.61	17.30	1.48	3.24	1.12	0.56	4.79	0.25	0.53
MWS-3	7.04	19.60	1.32	3.19	0.87	0.42	4.22	0.16	0.26
MWS-4	4.45	13.00	1.31	3.73	1.18	0.77	4.89	0.23	0.15
MWS-5	2.73	11.80	1.79	7.24	1.07	0.56	12.94	0.33	0.40
MWS-6	4.26	17.50	1.51	3.34	1.42	0.52	5.06	0.21	0.51
MWS-7	4.34	16.10	1.38	1.35	1.00	0.66	1.87	0.28	0.26
MWS-8	3.54	6.80	1.38	3.74	0.77	0.63	5.17	0.31	0.49
MWS-9	7.00	6.50	1.73	3.45	0.75	0.72	5.96	0.41	0.51

3 Results and Discussion

Table 4 Prioritized rank of micro-watersheds using the morphometric parameter

Micro-watersheds	R_b	Slope (%)	F	Dd	Dt	Rl	In	Rf	Rc	Compound rank	Priority
MWS-1	7	5	4	5	2	6	5	9	4	5.2	V
MWS-2	1	3	5	7	4	3	7	4	7	4.6	III
MWS-3	2	1	7	8	7	1	8	1	2	4.1	II
MWS-4	4	6	8	3	3	8	6	3	1	4.7	IV
MWS-5	9	7	1	1	5	3	1	7	3	4.1	II
MWS-6	6	2	3	6	1	2	4	2	6	3.6	I
MWS-7	5	4	6	9	6	5	9	5	2	5.7	VII
MWS-8	8	8	6	2	8	4	3	6	5	5.6	VI
MWS-9	3	9	2	4	9	7	2	8	6	5.6	VI

longer, potentially more stable watercourses, impacting drainage efficiency and sediment transport.

Secondly, the estimated stream length ratio for the Nileswar sub-watershed, as documented in Table 1 and ranging from 1.43 to 6.63, offers insights into the relationship between different stream orders. This ratio helps identify how the lengths of higher-order streams compare to those of lower-order streams.

3.2 Areal Properties of Drainage Basin

The area (A) and perimeter (P) of Nileswar sub-watershed are 190.46 km² and 78.28 km. It was observed through the attribute data management using a delineated boundary of watershed. Basin length (L_b) was calculated using a measuring tool in ArcGIS 10.8, and it was about 19.52 km. Basin width of the particular sub watershed was calculated using Eq. (2). It was obtained around 9.76 km.

The lemniscate (K) value of 0.5 for the specific sub-watershed indicates an important characteristic of its topography. A lemniscate value of 0.5 suggests that the watershed occupies a significant area within its origin regions and is associated with a relatively large number of higher-order streams. In hydrology and geographical analysis, a high K value indicates a watershed with a more expansive and interconnected drainage network, likely due to its complex topographical features. This information can be crucial for understanding the flow and distribution of water within the watershed and can have implications for land use planning and environmental management in the area.

Form factor is an important parameter to describe the shape of the catchment area. In this study area, the form factor value obtained for the whole sub watershed was about 0.5. The lower value indicated the elongated sub-watershed nature [11]. The smaller values of R_f indicated that the basin is relatively elongated and will have flatter peak flow over an extended time. Whereas the micro-watersheds with higher R_f value will have higher peak flow within short duration. In this study, it was found that the value of form factor varies 0.16–0.46, which depicts that the four micro watersheds (MWS-1, MWS-5, MWS-8 and MWS-9) is more circular as compared to other micro watersheds [9]. However, the MWS-3 will be generating the high peak in a short duration. The assigned ranking based on the form factor is summarized in Table 4.

3.2.1 Circulatory Ratio (R_c) and Elongation Ratio (R_l)

In this study area, the circulatory ratio was obtained through DEM management, which was about 0.4 for the whole Nileswar sub-watershed. Miller has described the basin of the circularity ratios range 0.4–0.5, which indicates strongly elongated and highly permeable homogenous geologic materials [12]. In this study circulatory ratio varies from 0.15 (MWS-4) to 0.53 (MWS-2). Strahler states that elongation ratio runs between 0.6 and 1.0 over a wide variety of climatic and geologic types. The varying slopes of watershed can be classified with the help of the index of elongation ratio, i.e., circular (0.9–0.10), oval (0.8–0.9), less elongated (0.7–0.8), elongated (0.5–0.7), and more elongated (<0.5). The elongation ratio of Nileswar sub-watershed is 0.79, which represents that the watershed is less elongated with high relief and high steep slopes [21] The MWS-3 is found as more elongated among nine micro-watersheds.

3.2.2 Rho Coefficient

Rho values of the Nileswar sub-watershed is 0.75. This is indicating higher hydrologic storage during floods and attenuation of effects of erosion during peak discharge.

3.2.3 Stream Density or Stream Frequency (F)

It is obtained from several streams and areas of the basin. Stream frequency of Nileswar sub-watershed was obtained about 1.53. The higher stream frequency values represent the watersheds with high resistant subsurface material and high relief [22], ultimately high runoff. In this study, MWS-5 produced more runoff as compared to other micro-watersheds [23].

3.2.4 Drainage Density (D_d) and Drainage Texture (D_t)

The drainage density relies on climate as well as physical characteristics, resistance to the erosion of the drainage basin. The calculated drainage density for Nileswar sub-watershed by using Spatial Analyst Tool in ArcGIS 10.8 was 3.49 km/km^2 indicating high drainage densities. It is indicated that the low drainage density indicates the sub-watershed is having highly resistant, permeable sub-soil and low relief [20]. Areas with thin vegetation are characterized by high D_d, which have large flood flow. The D_d values of the micro-watersheds vary between 1.35 (MWS-7) and 7.24 (MWS-5) (Table 3). The result indicates that MWS-7 has highly permeable sub soil material and is apt for ground water recharge.

High drainage density in MWS-5 depicts this region to be having high flood peaks and not suitable for cultivation unless adopting conservation measures [21].

Drainage texture is an important parameter in the field of geomorphology which gives idea about intensity of drainage characteristics. It measures closeness of the stream spacing. In the present study the drainage texture is obtained as 3.73 for Nileswar sub-watershed which indicates that the river basin has coarse drainage texture. All the micro-watersheds in the basin comes under coarse and very coarse category as the $D_t <$ 4, i.e., These will take longer duration to reach peak runoff [23].

3.2.5 Constant of Channel Maintenance (C) and Infiltration Number (In)

It is an inverse function of drainage density which gives an idea about the number of square kilometers required to develop a stream of one km. The computed value of constant channel maintenance for the river basin is 0.29 which indicates the soil with moderate erodibility.

Higher value of infiltration number indicates higher runoff, lower infiltration and vice versa. In this study, the infiltration number obtained was 5.33 which indicates the low infiltration. The MWS-5 has a higher infiltration number (Table 3), so that it is required to implement runoff control measures in this particular micro-watershed.

3.3 Relief Aspects of Channel Network

Slope is an important parameter to determine the morphometric characteristics of the catchment. The slope of the study area was derived from SRTM DEM using spatial analyst tools of 10.8 software. The slope range of this sub-watershed was estimated as 0–56.46° with mean slope of 8.2°. The slope map of Nileswar sub-watershed is shown in Fig. 7. The average spatial percentage slope corresponding to each micro-watershed is given in Table 3. The majority of the watershed (except MWS-8 and MWS-9) come under the strongly sloping category.

Aspect refers to the horizontal direction to which a slope of the surface faces. The aspect of the surface can significantly influence the local climate. In this study, the aspect map is derived from the SRTM DEM using Spatial Analyst tool in ArcGIS software. The output raster map shows the compass direction of the aspect with ranges from 0° to 360°. The aspect map of Nileswar sub-watershed is shown in Fig. 8. Maximum basin relief obtained was about 585 m. Relative relief of about 7.45 was obtained for Nileswar sub-watershed.

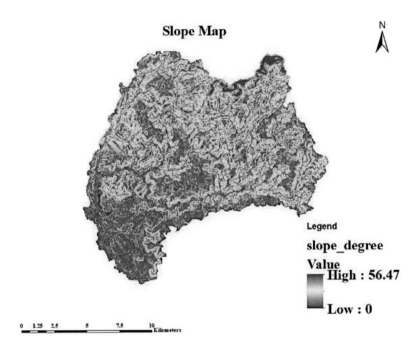

Fig. 7 Slope map of the Nileswar sub-watershed

3.4 Watershed Prioritization

Drainage characteristics of the basin relies on the relationship among the streams or rivers and slope, soil characteristics, rock resistance, structural and geological status of the basin. This study focuses on the prioritization of the Nileswar sub-watershed based on morphometric analysis. The compound score of total nine micro-watersheds and their ranking are given in Table 4. The prioritized score of micro-watersheds varies from 3.6 to 5.7 (Table 4).

High priority areas will be more susceptible to erosion. So that the study will help to identify more degradable land within a sub-watershed for conservation planning and implementation of conservation and management activities. In this study, morphometric parameters including bifurcation ratio, slope, stream frequency, drainage density, drainage texture, circularity ratio, form factor, elongation ratio and infiltration number were considered. Hence, the parameters like slope, bifurcation ratio, stream frequency, drainage density, drainage texture, infiltration number have a direct correlation with erodibility. Therefore, the prioritization was performed in such a way that the highest value of these parameters was assigned as rank 1and the smallest value has been assigned as last in rank. The other parameters including circularity ratio, form factor, elongation ratio have

3 Results and Discussion

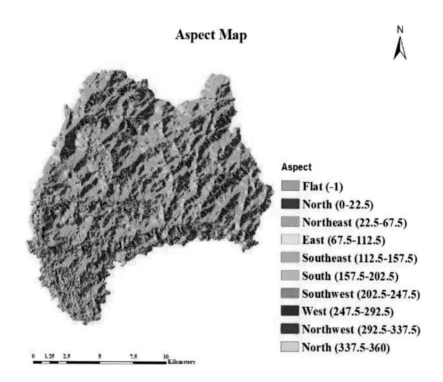

Fig. 8 Aspect map of the Nileswar

an inverse relationship with erodibility [22]. The least value of these parameters corresponds for soil with more erodibility. Hence, the least value was assigned as rank 1 and the highest value was marked as higher in rank [18] Finally, the compound parameter was calculated by taking the mean rank of all the parameters in particular micro-watersheds. The final priority was assigned in such a way that the least rating value is marked as highest priority; the next higher value was assigned second priority and so on [24, 25]. The sub-watersheds have been categorized into three categories as high (<4.7), medium (4.7–5.3) and low (>5.3) priority [21]. The micro watersheds MWS-2, MWS-3, MWS-4, MWS-5 and MWS-6 fall in the high priority classes, MWS-1 falls in the medium priority category and remaining micro-watersheds come under low priority classes (Table 4). MWS-6 should be given top priority while planning watershed management activities and should be focused to reduce runoff by adopting bunding and terracing practices.

The morphometric analysis is one of the most simple and proper methods with minimal input to quantify the problems regarding the land degradation, water distribution etc. In the present study, remote sensing and Geographical Information System (GIS) were utilized to derive morphometric parameters for the study area.

4 Conclusion

The geo-morphometric analyses of Nileswar sub-watershed were carried out through measurement of several linear, areal and relief aspects of the watershed with more than 25 morphometric parameters. The analysis of the stream network of the sub-watershed shows the dendritic patterns with coarse drainage texture. The variation in stream length ratio might be due to change in terrain characteristics such as slope, aspect and topography. The study found that the stream order varies from one to five. The drainage density values of the micro-watersheds vary between 1.35 (MWS-7) and 7.24 (MWS-5), which depicts MWS-7 has highly permeable sub soil material and is the potential zone for groundwater recharge. High drainage density in MWS-5 depicts this region to be having high flood peaks and not suitable for cultivation unless adopting conservation measures. The elongation ratio of sub-watershed is 0.79, indicating sub-watersheds fall under less elongated pattern.

The micro-watershed with the lowest compound rank was assigned as highest priority. The micro-watersheds were categorized into three classes as high (<4.7), medium (4.7–5.3) and low (>5.3) priority on the basis of their prioritized score. The micro watersheds MWS-2, MWS-3, MWS-4, MWS-5 and MWS-6 fall in the high priority classes among those MWS-6 should be given top priority while planning watershed management activities and should be focused to reduce runoff by adopting bunding and terracing practices as this area is a hilly terrain.

The study reveals that remotely sensed data (SRTM-DEM) and GIS technology in geo-morphometric analysis of drainage basin is more appropriate than the conventional methods. GIS based approach facilitates analysis of different morphometric parameters and to explore the relationship between the drainage morphometry and properties of landforms, soils and eroded lands. GIS techniques characterized by very accurate data management, mapping and measurement prove to be a competent tool in morphometric analysis.

References

1. Warjri, J.E. 2019. Estimation of soil erosion using revised universal soil loss equation in Upper Umiew catchment, East Khasi hills, Meghalaya, *Indian J. Soil. Cons.* 47(2): 156–162.
2. Horton, R., 1945. Erosional development of streams and their drainage basin; Hydro physical approach to quantitative morphology. Bulletin of the Gelogical society of America No. 56, pp 275–370.
3. Charizopoulos, N., Mourtzios, P., Psilovikos, T., Psilovikos, A. and Karamoutsou, L. 2019. Morphometric analysis of the drainage network of Samos Island (northern Aegean Sea): insights into tectonic control and flood hazards, *Comptes rendus-Geoscience* 351: 375–383.
4. Bal, S.K., Choudhuri, B.U., Sood, A., Mukherjee, J., and Singh, H. 2018. Geo-spatial analysis for assessment of agro-ecological suitability of alternate crops in Indian Punjab. *Indian J. Soil Cons.* 46(3): 283–292.

References

5. Kannan, R., Venkateswaran, S., VijayPrabhu, M. and Sankar, K. 2018. Drainage morphometric analysis of the Nagavathi watershed, Cauvery river basin in Dharmapuri district, Tamil Nadu, India using SRTM data and GIS. *Data in Brief*. 19: 2420–2426.
6. Strahler, A.N. 1964. Quantitative Geomorphology of Drainage Basin and Channel Network, Handbook of Applied Hydrology, pp 39–76.
7. Dash, C.J., Adhikary, P.P., Madhu, M., Maurya, U.K., Mishra, P.K., and Mukhopadhyay, S. 2019. Geospatial assessment and physical characterization of terraced low land (Jhola land) in Eastern Ghats highlands of India. *Indian J. Soil. Cons.* 47(2): 194–202.
8. Sreedevi, P.D., Sreekanth, P., Khan, D.H.H. and Ahmed, S. 2012. Drainage morphometry and its influence on hydrology in an semi-arid region: using SRTM data and GIS. *Environ. Earth Sci.* 70. 839–848.
9. Hejran, A.W. and Singh, K.K. 2017. Morphometric analysis of two major sub-basins of Helmand river, Afghanistan based on GIS approach, *J. Indian Water Res. Soc.* 37(3): 17–24.
10. Chorley, R.L. 1967. Models in Geomorphology. In: Models in Geography (Chorley, R.J. and Haggett, P., Eds.), London, pp 59–96.
11. Horton, R.E., 1932. Drainage basin characteristics. *Transactions, American Geophysical Union*, 13, pp 350–361.
12. Miller, V.C. 1935. A Quantitative Geomorphic Study of Drainage Basin Characteristics in the Clinch Mountain Area. Tech. Report-3, Columbia University, Dept. Geology.
13. Schumm, S.A. 1956. Evolution of Drainage Systems and Slopes in Badlands at Perth Anboy, New Jersey, Bulletin of the Geological Society of America, 67, pp 597–646.
14. Chethan, B.J. 2018. Influence of geomorphology on runoff characteristics of a catchment. M. Tech Thesis (Ag. Engg), Kerala Agricultural University, Thrissur, Kerala, India.
15. Faniran, A. 1968. The Index of drainage intensity-a provisional new drainage factor", *Australian J.Sci.* 31: 328–330.
16. Subramanya, K., 2015. Handbook of Engineering Hydrology fourth edition; Mc Graw Hill education 5:169.
17. Poongodi, R. and Venkateswaran, S. 2018. Prioritization of the micro-watersheds through morphometric analysis in the Vasishta Sub Basin of the Vellar River, Tamil Nadu using ASTER Digital Elevation Model (DEM)data. *Data in Brief* 20: 1353–1359.
18. Chandniha, S. K., and Kansal, M.L. 2017. Prioritization of sub-watersheds based on morphometric analysis using geospatial technique in Piperiya watershed, India. *Applied Water Science*, 7(1), 329–338.
19. Uhara, K.K.S., Ravikumar, V., Kannan, B. and Pannerselvan, S. 2022. Mapping the Potential Regions for the Construction of Cement Concrete Check Dams Using Remote Sensing and GIS. J Indian Soc Remote Sens 50, 2193–2208. https://doi.org/10.1007/s12524-022-01591-y.
20. Nag, S.K. 1998. Morphometric analysis using remote sensing techniques in the Chaka sub basin Purulla district, West Bengal. *J. Indian Soc. Rem. Sens.* 26: 69–76.
21. Gopinath, G., Nair, A.G., Ambili, G.K., and Swetha, T.V. 2016. Watershed prioritization based on morphometric analysis coupled with multi criteria decision making. *Arabian Journal of Geosciences*, 9(2), 1–17.
22. Javed, A., Khanday, M. Y. and Ahmed, R. 2009. Prioritization of sub-watershed based on Morphometric and Land use analysis using remote sensing and GIS techniques. *J. Indian Soc. Remote Sens.* 37: 261–274.
23. Mahala, A. 2019. The significance of morphometric analysis to understand the hydrological and morphological characteristics in two different morpho-climatic settings. *Applied Water Science*, 10(1), 1–16.

24. Mahmoodi, E., Azari, M. and Dastorani, M.T., 2023. Comparison of different objective weighting methods in a multi-criteria model for watershed prioritization for flood risk assessment using morphometric analysis. Journal of Flood Risk Management, 16(2), p e12894.
25. Shekar, P.R., Mathew, A., PS, A. and Gopi, V.P., 2023. Sub-watershed prioritization using morphometric analysis, principal component analysis, hypsometric analysis, land use/land cover analysis, and machine learning approaches in the Peddavagu River Basin, India. Journal of Water and Climate Change, 14(7), pp 2055–2084.

Printed in the United States
by Baker & Taylor Publisher Services